REMAKING THE NATION

Remaking the Nation presents new ways of thinking about the nation, nationalism and national identities.

Drawing links between popular culture and indigenous movements, issues of race and gender, and ideologies of national identity, the authors draw on their work in Latin America to illustrate their re-theorization of the policies of nationalism.

The societies of Latin America have placed great weight on nation building. Yet the nation is torn between the local and the global. Popular senses of affiliation to the nation are cross-cut by other bonds with communities and places; increasingly the countries of Latin America are experiencing the effects of globalization.

This engaging exploration of contemporary politics in a postmodern, post-new-world-order uncovers a map of future political organization, a world of pluri-nations and ethnicized identities in the ever-changing struggle for democracy.

Sarah Radcliffe is Lecturer in Geography at the University of Cambridge; **Sallie Westwood** is Professor in Sociology at the University of Leicester.

D1010189

REMAKING THE NATION

Place, identity and politics in Latin America

SARAH RADCLIFFE AND SALLIE WESTWOOD

London and New York

First published 1996
by Routledge
11 New Fetter Lane. London EC4P 4EE

Simultaneously published in the USA and Canada
by Routledge
29 West 35th Street. New York. NY 10001

© 1996 Sarah Radcliffe and Sallie Westwood

Typeset in Photina by
Keystroke. Jacaranda Lodge. Wolverhampton

Printed and bound in Great Britain by
Redwood Books. Trowbridge. Wiltshire

British Library Cataloguing in Publication Data
A catalogue record for this book is available from the British Library

Library of Congress Cataloguing in Publication Data
Radcliffe. Sarah A.
Remaking the nation: identity and politics in Latin America
Sarah Radcliffe and Sallie Westwood.
p. cm.
Includes bibliographical references and index.
1. National states. 2. Nationalism—Latin America.
3. Nationalism—Ecuador. 4. Racism—Latin America.
5. Racism—Ecuador. 6. Ethnocentrism—Latin America.
7. Ethnocentrism—Ecuador. 8. Multiculturalism—Latin
America. 9. Multiculturalism—Ecudaor.
I. Westwood. Sallie. II. Title.
JC311.R215   1996   95–26840
320.5′4′098—dc20   CIP

ISBN 0–415–12336–4
ISBN 0–415–12337–2 (pbk)

For Guy, Jessie and Dylan
S.R.

'Dedicated to the one I love'
S.W.

CONTENTS

FIGURES

ACKNOWLEDGEMENTS

It is a truism and a truth to say that a book results from the authors' interactions and shared reflections with a large number of people, individually and collectively. This book is no exception, and it is difficult to do full justice to the numerous friends, colleagues and acquaintances who between them encouraged us into producing the book.

In Ecuador, the fieldwork for the book was carried out in conjunction with a variety of people and institutions, who, along with the people who gave their time to us, deserve a large vote of thanks for their persistence, humour and dedication during the project. Nicola Murray, our knowledgeable researcher, was key to the success of the fieldwork, bringing her experience and insights, and innumerable routes through Ecuadorean *trámites* (bureaucracy). So to her, and her toddler Tuntiac, our heartfelt thanks. In Guayaquil and Esmeraldas, we wish to thank Norma Rodriguez, Gina Courzo Cangá, Pedro Castillo Cortés, Julio Arroyo Mina, Janet Preciado Camacho, Jackson Rodriquez G., Rosalit Olaya and Marlen Ponce Málala, and the Fundación para la Cultura Negra Ecuatoriana (FCUNE: Foundation for Black Ecuadorean Culture). In addition, Fidel Falconi, Jorge Salamon and Cecilia, and Ampam Karakras generously shared their knowledge and ideas, greatly enhancing our understanding of Ecuador. In Cotopaxi, the project would not have succeeded without the assistance and commitment of Lourdes Llasag and members of her family, Angel, Rebecca and Raul. Elsewhere, Gaby Costa was an insightful museum guide. In Quito, the research centre FLACSO (Facultad Latino Americana de Ciencias Sociales) provided us a stimulating intellectual environment; we would like to thank Amparo Menendez Carrión, Francisco Carrión, Adrian Bonilla and Fernando Bustamante. In Colombia at the Universidad del Valle in Cali, Gabriella Castellanos and Miryan Zúñiga gave generous and enthusiastic support to the project.

In Britain, we have been encouraged in writing and fieldwork by a wide range of individuals and institutions. The Economic and Social Research Council provided us with a grant for fieldwork and research assistance (ESRC grant no. R000234321). Thanks are also due to the Geography Department, Royal Holloway, University of London, for a sabbatical allowing time to think, read and write, and to the Department of Sociology, University of Leicester, and the Institute of Latin American Studies, London. Especial thanks to Felix Driver, David Gilbert, Denis Cosgrove, Klaus Dodds and Tim Unwin at Royal

Holloway, as well as James Fulcher, Vesna Popovski and Annie Phizacklea at University of Leicester. John Dickenson, Jane Jacobs, Cathy Boyle, Tessa Cubbitt and Pete Wade made comments on draft chapters, while Robert Ash worked on the bibliography. Anthony Giddens was unfailing in his support, and David Lehmann encouraged us to talk to Anne-Christine Taylor and Nicola Murray. David Preston had literature and ideas to share. Hannah Bryan-Brown, Natasha Middleton and Miguel Tuscano all facilitated the project at various points, for which we are thankful. Justin Jacyno at Royal Holloway and Pam Spoerry at Cambridge University both saved the day with their technical skills at the last minute, so many thanks. In the last year, the writing-up in Leicester would not have been possible without the generous support of Rashpal Singh, Tirathpal Singh Naute and Rakesh Patel who, each in their own way, have contributed so much to the production of this book, warmest thanks to you all.

We are grateful for permission to reprint material amended from two articles by Sarah Radcliffe, previously appearing in *Ecumene* (3(1) 1996, Edward Arnold), and in *Gender, Place and Culture* (3(1) 1996, Carfax Publishers, PO Box 25, Abingdon, Oxford OX14 3UE).

Sarah Radcliffe and Sallie Westwood
Cambridge and Leicester 1996

GLOSSARY

abertura (Portuguese)/*apertura* (Spanish) literally, 'opening', referring to the process of political opening occurring after the bureaucratic authoritarian regimes of the 1970s and 1980s.

altiplano highland plateau in the Andean countries.

atrasado/a 'backward', with connotations of 'uncivilized' when applied to certain groups.

apu deity embodied in mountain peaks in Andean zone.

audiencia colonial administrative unit, with judicial powers.

autolinderación name given to the drawing of land-title boundaries by indigenous groups.

barrio an urban neighbourhood, specifically a low-income area.

blanco/a a person with 'white' phenotypical characteristics, or European heritage.

blanqueamiento ideology and practice of 'whitening', or becoming more 'Europeanized'.

calle street, or in certain circumstances, the public sphere.

campesino/a peasant or rural dweller; term used in Andean countries increasingly after the 1970s in order to avoid the use of the more racialized term of *indio* or *indígena*.

casa house and domestic space.

CEBs – comunidades eclesiasticas de base literally Christian base communities. the informal church-formed groups whose ideological base is founded in Liberation Theology.

chicha music: a form of musical expression which emerged in Lima, Peru, during the 1970s and 1980s out of the synthesis of Andean (e.g. *huaynos*) and coastal (e.g. *marinera*) forms.

cholo/a a loaded, and often derogatory, term for people of presumed rural (and indigenous) origin who have adopted more urban and/or *mestizo* styles of dress, speech and social practices.

choloficación process of becoming a *cholo/a*.

colonos colonizers or settlers coming into the Amazonian region of the Andean countries.

compradito a male personage in the tango dances of Argentina.

CONAIE Confederación de Nacionalidades Indígenas de Ecuador Confederation of Indigenous Nationalities of Ecuador.

costeño person from the coastal region of the Andean countries.

costumbrismo a type of Latin American literature of customs and manners, in the nineteenth and early twentieth centuries.

creole during the colonial period, a person of Latin American birth and European parentage; now the term 'creole elite' refers to people of 'white', largely wealthy, background.

desaparecido/a literally, disappeared person; victim of military regimes' campaign of arbitrary arrest and murder.

dicho saying.

ecuadorianidad 'Ecuadorian-ness', or a feeling of authenticity associated with the country of Ecuador.

empleada houseworker or domestic worker.

FCUNE *Fundación para la Cultura Negra Ecuatoriana* Foundation for Black Ecuadorean culture.

fotonovela cartoon-style magazine with picture-based narrative form.

fútbol football, soccer.

gaucho cowboy, herdsman, particularly associated with Argentina.

hacendado estate owner.

hacienda rural estate.

historieta strip cartoon.

huasipunguero/a tied labour contract in rural estates.

indigenismo ideology and political policies associated with the revaluation of indigenous culture, particularly during the 1920s and 1930s in Peru and Mexico (associated with such writers as Mariátegui and Vasconcelos); now also associated with the term *neo-indigenismo* in Andean government policies on culture.

indigenista person or writing advocating the position of indigenism.

indio/a a person of indigenous origin, term can be used in racisms to denote a person of inferior status.

latifundista owner of large agricultural estate.

lugar place in geographic sense, and also, in certain circumstances, place as rank or order.

lugareño/a person in a place, especially a rural location.

machismo ideology and practice of male domination over women.

maquiladora or *maquila* export-oriented production facilities.

marianismo ideology and practice of female behaviour modelled on the Virgin Mary or María.

mestizo/a a person with both Spanish (or European) and indigenous heritage.

mestizaje the ideology and practice of creating new 'races', through miscegenation; under the myth of *mestizaje*, the majority of the national population were to be mestizos and become the new Latin America.

montuvio/a (or *montubio/a*) person of mixed, including black, heritage.

mulatto a person of both African and Latin American descent.

neo-indigenista associated with reinterpretation of classic indigenist writings in 1980s and 1990s.

niños children, but also applied to groups considered 'childlike'.

pampas plains or grasslands.

patria nation, 'fatherland', or homeland.

patria chica literally 'small homeland', meaningful place, smaller than nation, to which one has emotional attachment, as in place of origin or urban neighbourhood; term particularly used in Peru.

patria potestad legislation providing men certain powers over wives and household members, from work to marital relations.

peronistas followers of Juan Perón, President of Argentina.

pintura de tigua form of painting on hide found in Ecuador, using naive styles and acrylic paints.

poblador resident in a *barrio* or neighbourhood, from *población* (settlement).

lo popular the products or sentiments associated with the 'people' or the low-income population of nations.

PRI Partido Revolucionario Institucional Institutional Revolutionary Party of Mexico.

el pueblo the 'people' of a nation.

quechua/quichua indigenous people and language. Quechua is spoken throughout the Andean countries by some 9 million people; the spelling *quichua* is used in Ecuador, *quechua* in Peru and Bolivia.

raza literally 'race', but with connotations of white-European inheritance.

serrano/a person from the Andean region of Ecuador, Peru or Bolivia; also used to describe an aspect of Andean culture – reserved and distant in relationships.

suca Quichua term connoting beauty, associated with white facial characteristics.

telenovelas soap operas, transmitted on television.

trago sugar-cane alcohol.

traje clothing associated with indigenous population in Guatemala, and appropriated as a symbol of national identity.

vecino/a neighbour.

INTRODUCTION

Remaking the Nation is an exploratory intervention into the complex terrain of nations, nationalisms and national identities in Latin America. While our focus is regionally specific our concerns are prompted by a series of pressing political and theoretical issues that simultaneously threaten and sometimes disrupt the fragile democratic settlements of earlier periods. We live currently in a world suffused with nationalist discourses and yet more securely woven together via international capitalism than ever before. Our age is beset with contradictions, some of which are condensed into nationalist fervour. In entering this terrain, we do so with some trepidation, offering an analysis which we hope will encourage debate and, for those unfamiliar with Latin America, will inform in ways that stimulate a rethinking of global concerns.

Both the research presented here and the book grew out of our earlier collaboration on the edited collection, *Viva* (Radcliffe and Westwood 1993), a series of studies of gender, racism and popular protest which elaborated a theorization of both the externalities and the subjective processes of political action in the context of the 'new' democracies of Latin America. Using a Foucauldian account of 'sites', power and the body, the book generated a series of substantive accounts of power struggles, identities and locations as a way of reconstructing our understandings of popular protest and its gendered and racialized forms. The work on this book prompted a new series of questions concerned with the developing political agenda in Latin America which, as elsewhere, foregrounded the nation and national identities in innovative ways.

The rise of the nation-state and the construction of national identities in the modern period were key moments for Latin America, where indigenous and European identities have been profoundly transformed during five hundred years of political, cultural and social change. Currently, these societies, in which the military have played a crucial role, are once again in the process of re-formation through the development of discourses with claims to indigenous and diasporic identities. Part of the struggles for democracy in Latin America, these claims to nationhood and new national identities are being generated from within and beyond the boundaries of the old states. The post-modern world of Latin America offers a vision of pluri-nations and ethnicized identities which is of crucial relevance for all parts of the world.

We have conducted the research and written this book at a specific juncture in the history of the social sciences, a moment at which discipline boundaries are increasingly broken down and reconstituted in novel and exciting ways. We are a product of this, coming together with intellectual biographies fashioned from geography and sociology individually while sharing a background in social anthropology. We also share the notion that subjects and disciplines are not neutral knowledges but political understandings in every sense of the word political. The production of social science knowledges is a political act and we are conscious of the multiple positions that we occupy, as white middle-class women, as researchers and academics. But we do not accept that our understandings cannot be larger than this. Consistent with our theoretical frame we recognize that we cannot stand outside discourses; we can stand within many and generate a multiplicity of insights that will, we hope, assist in the analysis and theorization of national identities in Latin America.

In part our research grew from our previous collaboration, the politics of the Latin American states, but also from a degree of disquiet, a discomfort with an ever expanding literature on nations, nationalism and national identities influenced by post-structuralism which, while often exciting and illuminating, privileged theoretical concerns and was increasingly textual in its emphasis. We have learned an enormous amount from this corpus of work but we were still left with a series of questions about the power of national sentiments and the potency of national affiliations in people's lives, for which we could not provide satisfactory answers or explanations. Many of these questions, of course, remain unanswered and continue to preoccupy our thoughts. We decided, in effect, to shift the focus from literary and artistic productions as texts to an initial anthropological concern with social life as text, returning to Geertz (1973) and the notion of 'thick description' as a way of trying to answer our questions, most especially the issue of the saliency of national identities. National identities were, it seemed to us, everywhere and nowhere, represented in maps, anthems and flags, football matches and presidents. But this tangibility made opaque the nebulous quality of the construction we sought to theorize and analyse. We concluded that national identities needed to be thought in more substantive ways and that a study that began with this premise should be attempted. Most simply, we took Anderson's (1991) 'imagined community' and asked how is it generated, sustained and fractured in one nation-state – Ecuador.

The research was framed by the understanding that both 'official' and 'popular' nationalisms contribute towards the generation and sustenance of the imagined community of the nation. However, the distinction between 'official' and 'popular' does not signal a binary opposition but a complex articulation which both supports and fractures the nation. Equally central to the research frame was the understanding that identities are multiple and that the de-centred self produces a series of complex relationships to the nation. De-centred selves, however, are not endlessly fragmentary but constituted in relation to biography, history, culture and, most importantly, place. Importantly, these 'positionalities' are also the moments of a de-centred nation which offer a variety of distinctive relationships to 'the nation' and the constitution of the national. In order, therefore, to better understand these relationships – and the interaction between the fractured social and the de-centred subject – of the national story, the study was organized around these fractures. Consequently, the research began with the importance of place and seven distinctive sites in Ecuador were chosen in which to conduct fieldwork, ranging from the main urban centres (Quito and Guayaquil) to rural areas, small towns, Amazonian villages and the coastal city of Esmeraldas, the latter famous for its long history of black settlement and its proximity to Colombia.

Place as a defining marker of the national experience and, therefore, of the lived experience and understanding of a national identity is not intended to imply inert geography and the notion of region, but the political, economic, cultural and social dynamics of place. Consequently, the ethnicization and racialization of space was a key element in all these sites. In addition, the economic relations of the nation-state were built into the samples which included the wealthiest sections of the main cities and the poorest *barrios* (low-income neighbourhoods), both those with long-term settlement and more recent migrants to the city. In Quito, for example, we interviewed both the elite of the capital city, and people in a working-class neighbourhood within the old city as it is now a conservation project, including many of the buildings and functionaries that symbolize the nation-state.

In all seven sites we interviewed one hundred subjects, women and men equally in the hope that this would unpack the complexities of the gendering of national identities (although this has proved to be among the most opaque). Thus, during 1993 and 1994, we interviewed Ecuadoreans across class, gender, ethnicity and age groups, producing seven hundred responses from a complex interview schedule. There were also a large number of semi-structured interviews with members of the military, politicians, priests, educators, civil servants, members of non-governmental organizations (NGOs), women's groups, community groups and the leaders and members of the Confederation of Indigenous Nationalities of Ecuador (CONAIE) and the main African-Ecuadorean organization, FCUNE. Some part, but by no means all, of this material has been used in the production of this book.

Having decided on a research frame, we were still faced with the task of relating the 'externalities' of the nation – that is, forms of governance, national histories, the flags, icons, and so on – to the 'internalities', the subjective experience of a national identity within the complexities of place, class, gender and age. In order to explore this, we constructed a complex interview schedule in Spanish (also translated into Quichua for use in one Andean area). The interview schedule gathered information on the class, educational and occupational backgrounds of respondents, and then sought ways in which to draw out the imaginaries that interviewees constructed of the nation-state of Ecuador by means of questions on the ethnic composition of the country, the national and local power structure. In addition, we asked people to identify the flag, the map of Ecuador (Figure I.1) and, importantly, everyone we interviewed in Ecuador could do this. We also asked people to name the most beautiful place in Ecuador, what they thought symbolized Ecuador in the world and so on. The complexities of this interview schedule meant that

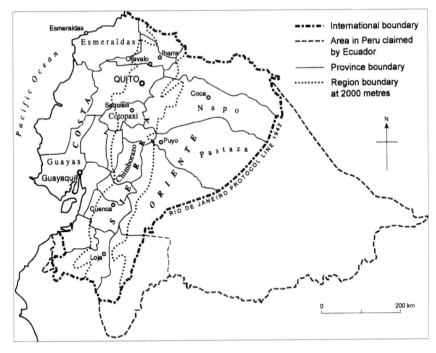

Figure I.1 Map of Ecuador

interviews were long, especially in relation to the numbers of questions without pre-coded responses. The study overall was, therefore, layered with complexity; this is no surprise in a study that sought to answer the question: How is the imaginary of the nation discursively constructed? Choosing a small country such as Ecuador made it possible to work with this complexity, but it should not be understood as a 'representative' study, rather as a case study which, we hope, illuminates the processes of nationalisms and nation-building within which national identities are formed. It is our hope that it could, nevertheless, offer theoretical and analytical frames that can be of use elsewhere in Latin America. Thus, in writing this book we have not sought to present a monograph on Ecuador but a book which draws upon both the material from the Ecuador study and extant materials on Latin America more generally.

Ecuador embarked on a self-conscious nation-building process following independence from Gran Colombia in 1830. 'National' identities emerged only slowly over the next hundred years, a product of a constant 'mirror dance' between regional, subnational and local affiliations. Well into the republican era, three regions – based around the major cities of Quito, Cuenca and Guayaquil – had distinctive cultures and a high degree of autonomy, politically and economically (Quintero Lopez 1987).

Although no longer under colonial rule the administrative separation of 'Indians' and 'Spaniards' continued into the republic, disregarding a high degree of miscegenation. The imagined superiority of the cultured *raza* or race of Spaniards and their descendants rested upon the pre-eminence of the Spanish language and religion, and a shared sense of commonality with the other elite creole (Spanish born in Latin America) groups in neighbouring countries. Populations racialized through the use of terms such as indians, black, *cholos* (urbanized indigenous groups), *montuvios* (people of mixed, including black, heritage) and *mestizos* (Spanish-indian populations) were excluded from national life during the early republican period. Indeed, the elite's dominance depended in part upon the reproduction of their difference from indigenous and mixed groups (Deler and Saint-Geours 1986: 428).

That the centre of the nation did not hold was revealed by border disputes with Colombia and Peru in mid-century, when the state's incapacity to protect the 'conditions for the production of the nation' (Quintero Lopez 1987: 134) was exposed. 'Foreign' powers were asked for military support in settlements of 'internal' disputes, while private armies were maintained by landlords. The crisis peaked in 1859 when five regional governments each claimed to be 'national'. Threatened with invasion by Peru, the regional powers of Quito, Cuenca and Guayaquil united and at the same time carried out some 'national' reforms, in effect beginning a process of nation-building.

The indigenous tax was abolished, along with enslavement for blacks; landowners, while maintaining their power in production in a semi-feudal economy, increasingly worked in a 'national' legal and administrative structure. Between 1860 and 1894, a *latifundista* (large estate owner) state was consolidated through judicial and religious-ideological means and labour relations, although some regional fragmentation of power remained. Bourgeois representation was limited, with only 5 per cent of the population voting before 1883. Some coalescence did occur around the notion of 'Ecuador' with the emergence of key integrating political figures such as Eloy Alfaro, García Moreno and Rocafuerte (Deler and Saint-Geours 1986).

However, only during the 1880s and 1890s – culminating in the 1895 revolution – did the concepts of a 'national' interest come to the fore. The idea of the state as arbiter between equal citizens and protector through legal provisions emerged with the Liberal revolution (Clark 1994). As centralized tax collection began and a national bank was formed 'for the public interest', so the state apparatuses and the legal system came increasingly to represent a national community. In 1885 the national currency, the *sucre*, was introduced: previously the money of neighbouring countries was used. Peasants were incorporated into military service and state secular education was introduced (Quintero and Silva 1991). During the last decades of the century communications were improved via the telegraph (introduced in the 1880s) and the Quito–Guayaquil railway (completed in 1908), while provinces adjoining Peru became more integrated into the Guayaquil region. The features of nation-building introduced at this time, such as military service, state education and state bureaucracy, all typify Anderson's 'official nationalism' developing in Europe during the same period (Anderson 1991)

Yet the emergence of the more centralized 'nation' did not mean the disappearence of local and regional differences. While the Sierra was influenced by the Catholic Church in its rituals and rhythm of life, the Costa (Coast) was characterized by a weaker Church challenged by liberal newspapers, workers' associations and other groups. Already by 1880 most schools in Guayaquil were secular rather than religious. Overall, the imagining of the nation became an increasingly secular affair consistent with the key liberal idea: the separation between Church and state. In 1902 civil marriage and divorce were introduced, followed in 1908 by the nationalization of the properties of religious orders (Clark 1994).

As elsewhere in the Andes, indigenous and black groups remained largely 'unimagined' in the national community despite the 'Liberal' Revolution. Citizenship rights were not extended to these marginal groups, although the worst excesses of forced labour were curbed. In an effort to gain a national role, *Serrano* (highland) indigenous groups organized numerous uprisings against the state and landowners. Between 1913 and the mid-1920s, indigenous groups in the provinces from Cotopaxi to Azuay protested at censuses and forms of labour recruitment in an effort to redefine their relations with the landowners and the developing state structures. Concurrently, there was the development of an intellectual debate about the 'indigenous question' similar to the *indigenista* (pro-indigenous) movements in Peru and Mexico.

By 1942 the population of Ecuador was 3 million: indigenous groups made up approximately 39 per cent while peasant groups accounted for 65 per cent; 56 per cent of the population were literate (Quintero and Silva 1991: 313). In 1960 the estimate of the population remained the same '3 million' but the estimate of the indigenous population had risen to 50 per cent. Many were tied to Sierra *haciendas* (rural estates) through *huasipunguero* (tied labour contracts) arrrangements, in which a plot of land was given in return for regular full-time work (with minimal or no wages) on the *hacienda*. Other family members also worked as domestics and labourers. Not surprisingly, literacy rates were low with rural literacy reaching 54 per cent among men, and only 36 per cent among women (Andean Mission of the United Nations 1960). Still perceived as outside the nation, the indigenous populations were subjected – with United Nations support – to national programmes of integration in the 1950s and 1960s.

In the 1990s Ecuador continues to grapple with the task of nation-formation. With a population of around 10 million (Ecuador 1991), and a decade-long programme of neo-liberal policies, the country has experienced de-centring through globalization and the need to provide a unifying account of the nation-building project. The border dispute with Peru provides an occasional focus for national identities (particularly in January 1995), although strong regional affiliations continue to cross-cut Ecuadorean nationalism. With a 55 per cent urban population and 38 per cent under the age of 15, the concentration of subjects within the sites closely associated with the official project of nationalism (urban locations, educational establishments) suggests the possibility for imagining the nation (89.9 per cent literacy rates reinforce this impression). Yet the coexistence of diverse regional and ethnic cultures (the most frequently given figures for ethnic makeup is indigenous groups 40 per cent, *mestizos* 40 per cent, whites 15 per cent and blacks 5 per cent) and the contestations of what have historically been 'white', urban and elite conceptions of the nation all indicate the fractures within Ecuadorean national imaginaries.

While Ecuador cannot be fully representative of the region, its history and contemporary social dynamics can be taken as illustrative of processes elsewhere and we have used the case of Ecuador in order to develop a theorization and analytical framework which is elaborated in the chapters that follow.

The book begins with a theoretical overview which provides a framework for subsequent chapters and incorporates our understandings of the key sites in which national identities are generated and sustained. The first of these sites is the imaginary and the ways in which imagining nations provide the context within which national identities are called forth. The second, articulated with the first, is the body as a site for the play of powers so crucial to the management, disciplining and identification processes of nationalisms and national identities. But imaginary and embodied nations are lived through the discursive practices of everyday life elaborated in the popular in what we have called living nations. Finally, our concerns with the spatial, from the power of maps to symbolize the nation to the sense of place offered by neighbourhoods and localities, are elaborated in our discussion, placing the nation. These major axes around which we have worked are not a series of hierarchical levels but a set of articulations which both generate a fragile consensus – a centred nation – and equally fracture the nation, contributing to the de-centring of the national. While this provides the framework within which we are able to explore the saliency and variety of relationships to the nation and the complexities of national identities, we have also tried to find ways of theorizing the processes which interpellate subjects and explain the power of attachment or, in another language, the investments made by people in their national identities. For these processes we have used two terms: first, 'geographies of identities' which clearly emphasizes the saliency of place, and second, 'correlative imaginaries' which provides a vocabulary for discussing the seductions of nation, nationalism and national identities. This framework and the new vocabulary we have used are presented in Chapter 1.

Chapter 2 elaborates and then deconstructs the imaginary of a fictive ethnicity around which nation-states, including Ecuador, are organized. Discussing the importance of racism in the Latin American states, the ways in which it fractures and reorganizes the national project, thereby de-centring the nation, is examined. In Chapter 3 we return more centrally to Ecuador as a case study of the development of an official nationalism, bound to the importance of geography, alongside the re-presentation of the nation, its history and population. Chapter 4 examines the living nation through the processes of reconversion and the ongoing articulation between the official and the popular. In Chapter 5 we foreground place and the importance of land and locality in national imaginings, introducing the variety of ways in which the spatial saturates everyday lives. Our concerns with the embodied nation examined in part in Chaper 2 in relation to racism are again privileged in Chapter 6, which focuses attention upon gender and sexualities, teasing out the complexities of the reorganization and representations of gender and sexualities in relation to the national imaginary. Finally, in Chapter 7 we return to the key questions of sentiments, affiliations and the affective relations of national identity often expressed in a sense of belonging. It is this sense of belonging which occupies us in the closing pages of the book. In concluding we suggest that it is in the national identities of complex multi-ethnic, regionally divided, post-colonial states of Latin America that it is possible to glimpse a politics that contributes to cross-national imaginaries. In the conception of pluri-nations, multiple selves and multi-ethnic identities can have a sense of place and belonging countering a privileged basis for national identity.

1

IMAGINING THE NATION

Rethinking national identities

This is a book about nations and national identities in the late twentieth century. It is organized around four major themes: the imagining of the nation; the embodying of the nation; living in the nation; and placing the nation. Through thinking about the nation in these ways, we arrive at a reconceptualization of the nation and its everyday ramifications.

IMAGINING THE NATION

Broadly speaking, we can recognize three main models for Latin American nationhood, undertaken by civilian or military regimes: intellectual proposals, military projects and political projects. Within the intellectual proposals have been several radical alternatives, elaborated conceptually rather than in practice, including the Latin American communist ideas, Haya de la Torre's ideas, and finally Mariátegui's notions. In the 1920s and 1930s, the communist idea influenced many Latin American intellectuals, persuading them of a need for separate indigenous nations with their own rights, for example the Andean Quechua and Aymara. While such proposals would appear to be anachronistic today, their programmes have not been forgotten entirely; in our interviews with elites in Ecuador, one elderly man suggested (wrongly) that the contemporary Ecuadorean indigenous movement was a strategy to realize the 1930s communist proposals. Emerging at the same time as the communist proposals were other projects, such as those of Victor Haya de la Torre and José Carlos Mariátegui. A Peruvian political thinker, Haya de la Torre proposed a populist alliance between the working class and the national bourgeoisie, disregarding the significant peasant and more indigenous sector, especially in the Andes. Another Peruvian writer, Mariátegui, proposed integrating indigenous groups into national society to make one nation by overcoming social, ethnic and economic divisions (Rowe and Schelling 1991).

Military or authoritarian projects for nationhood have different rationales and proponents in Latin America. Important politically since the Independence wars of the early nineteenth century, the region's armed forces have often felt justified in intervening in society to carry out measures aimed at security, development and nation-building (Rouquié 1987). Drawing upon a rationale which highlights the political and social stability that military regimes provide, projects involving the armed forces have been highly diverse, at times oriented towards the forced homogenization of the citizenry (by eliminating people whose political or ethnic affiliations appear to challenge the realization of nationhood, such as in 1970s Argentina and 1980s Guatemala). In other countries, the armed forces have often been characterized by their separation from civilian society, creating and maintaining their own institutions (schools, colleges, housing), their own economic interests (companies, banks) and their own social networks. While conscription into the military provides for some a route into politicized nationhood, its importance is reduced by patchy takeup, and the large gulf between professional military personnel and conscripts.

The political projects for nationhood – the third major group – have been generated by both military and civilian regimes in recent Latin American history, such as under the military junta in Peru between 1968 and 1975, or Argentina under the *peronistas* (followers of the Peróns). Utilizing corporatist measures in an attempt to mobilize citizens from the top down', the Velasco regime in Peru attempted to create a less socially and ethnically divided country. Whether under its authoritarian or corporatist guises, populism has been perhaps the most widely known of the political projects for nation-hood, referring to the articulation of 'the people' into political regimes, with Peronism in Argentina perhaps the best known case well illustrating its force and persistence. Gathering up diverse groups into an imagined national community around the ideas of *el pueblo* (the people) and *lo popular* (the people's cultural and social resources), populism has had numerous and varied adherents throughout twentieth-century Latin American history (Laclau 1977). The persistence of populism and the existence of numerous other models for nation-building do not mean that the 'huge ideological work' of national identity is already done or ready-made. Rather, the process of defining the 'people' in a particular nation is a long-term one (Greenfield 1992), and often conflictual, demanding ideological work to create homogeneity with*in* the nation and distinction from what is *outside* the nation.

> We are not Europeans, we are not Indians, but a middle group between the indigenous and the Spanish, Americans by birth and Europeans by right.
> (Simón Bolívar, Independence leader in the early nineteenth century, quoted in Brading 1991)

Analysing the processes of nation-building, Latin American histories point to several key features which distinguish the region from other post-colonial countries. With its early colonization and early independence, Latin American countries have a long history of post-coloniality and nation-building projects, marking differences in timing and context in comparison with Europe, Asia and Africa. The history of Latin America – contrasted with the longer history of indigenous Abya Yala (Platt 1992) – erupted violently with the arrival of Spanish and later Portuguese conquerors from 1492 onwards. Nineteenth-century anti-colonial struggles were heralded by the indigenous uprisings of Tupac Amaru and Tupac Katari in the late eighteenth century. By the early nineteenth century, the 'creole pioneers' of nation-building based their search for nationhood on the notion of a common people, a community united in its political independence from Spain and Portugal. Latin American nations thus have a long history, comparable to many European states.

While independence leaders 'imagined' all Peruvians or Mexicans as equal citizens, in practice social and ethnic divisions gave widely differing experiences for whites, blacks and indians under the creole nationalisms between 1770 and 1830 (Brading 1991). From the colonial period, the Americas provided labour and capital for European capitalist development, and indigenous and black labour had ties with Iberia, Italy and the Low Countries (Blaut 1987). Colonial relations brought peninsular Spaniards into the region, just as capitalist relations brought Asian and European populations after independence. When black slavery was abolished from the mid-nineteenth century, other labouring populations (Japanese, Chinese and southern European) were drawn into economic circuits and labour markets, providing the basis for peripheral supply of primary goods to the metropoles.

However, sustaining the imagined communities of these newly developing nations is the result of complex relationships between representations, subjects, the media and identity. Anderson (1991) was the first writer to highlight the importance of the 'emergence of print capitalism' for the creation and consolidation of national identities. By making available – through newspapers and novels – a knowledge of shared or known activities, he argued, there emerged a sense of an 'imagined community' of the nation among an increasingly literate population.[1] The coming-into-being of the nation as 'cultural signification' (Bhabha 1990a) in the early days of nationhood, such as among the creole pioneers, was certainly linked with print capitalism (Anderson 1991; Brading 1991; cf. Colley 1986 on Britain).

In certain circumstances, literature provides images and knowledge of a national history and geography, a 'gallery of representations' (Poole 1992: 16). After independence, allegorical novels were explicitly expected to fill the 'gaps' in national histories, creating imaginary linkages across regional, economic and ethnic lines. In the nineteenth century, a widely circulated literature of manners or *costumbrismo* offered the possibility of mutual comprehension to different social strata, even if it did not make them into a horizontal community of equals. Such 'national romances' were to provide local narratives, 'writing *América*' rather than Latin America-as-seen-by-Europe (Sommer 1991: 10).

The representations on offer in novels, newspapers and other media come to be read and experienced as 'commonsense' and thus gain hegemony, creating what is experienced as an adequate and unalienated representation of subjects' lives. Subjects can thereby abstract certain images from their own lives which then 'stand in' for the nation:

> images are then projected onto the generalizing screen of the 'national imaginary' as fetishes of the nation which stand in for the thing itself.
> (Bowman 1994: 142)

However, as the type and number of media channels (radio, television, film and so on) proliferate in late capitalism, so too do the means by which imagined communities can be created and contested (cf. Schlesinger 1987). In Mexico after the national revolution in 1910–17, film – and not writing – was the medium through which national identity was articulated and consolidated (King 1990). The new media can at times be used by the state in the propagation of official nationalism, as in the case of the use of television by the Brazilian authoritarian military regime between 1964 and the mid-1980s (Rowe and Schelling 1991). Increasing international control and circulation of information and images complicate any idea of a bounded national community with its own communicative networks and specific cultural forms. In Latin America, the reception of information from 'international' Western news agencies has concerned some Latin Americans about the ways in which they are re-presented to themselves (Mattelart 1979). However, even without Western media, the ironies and contradictions of (non)national media can be seen. For example, in the Amazonian Oriente region of Ecuador, many people cannot receive national television broadcasts, although reception of Peruvian programmes is good. Viewers feel guilty watching 'another' television, feeling that their national affiliation is being questioned.

Moreover, with the presence of large functionally illiterate or semi-literate populations in much of the developing south, the emphasis on written forms of culture and other 'high' cultural artefacts is, perhaps misguidedly, focusing attention on the preserve of a literate minority. Pointing to the use of oral transmission of narratives and iconography in Latin American popular nationalisms, Rowe and Schelling argue that 'the weakness of Anderson's scheme lies precisely in its omission of the role of popular culture' (Rowe and Schelling 1991: 25). Among one indigenous group in Colombia for example, indigenous identity and difference is represented *visually* in the form of a flag which incorporates male and female symbols as well as national references. Romances of identity can thus incorporate and utilize *non-narrative* forms (Rappaport 1992). The widespread use of national calendars, holidays, festivals and uniforms all link into the non-narrative but highly visual 'liturgies' of nationalisms (Mosse 1975).

While Simón Bolívar, a famous independence leader in the 1820s, could claim that 'we are Americans by birth' the debate about the origins of Latin American subjects has continued to shape nationalist discourses until today. The exact nature of interaction between indigenous and Hispanic legacies is an enduring question in discourses about identity, making the name and the nature of the region highly contested and political (e.g. Platt 1993; Mallon 1995; Thurner 1995). The 500 years of resistance campaign of the early 1990s was one specific outcome of this politics. Challenging the prevailing interpretations of the conquest as worthy of celebration – on the quincentenary of European arrival in the Americas in 1992 – the campaign united subaltern indigenous and black views on colonialism and post-colonialism. Drawing attention to the suppression and disappearance of indigenous cultures, enslavement of black populations, and removal of political and personal rights, the campaign revealed the numerous contradictions and divisions within the post-colonial project (*History Workshop Journal* 1992; *Race and Class* 1992).

The existence of multi-ethnic, multi-diasporic nations raises particular issues in the analysis of national identity. In societies or cultures shaped by the experiences of colonialism, mass migration from abroad or the co-existence of various ethnic groups – characteristics that include the great majority of present nations – the question of simultaneous yet different versions of national identity is of great significance. In colonized societies, especially in the south, the imagining of the nation took place in the context of the colonizers' representations of the national community. In this sense, colonized nations had to imagine difference and alternative national identities from the start (Chatterjee 1993: 5). Given the early independence and long post-coloniality of Latin American nations, the formulation of national identity lies in the diversity of responses to, and uses of, the modern forms of the nation.

Latin American writers emphasize their countries' socio-economic diversity and the difficulty of forming a national community from such heterogeneity. The work involved in national self-definition is not, according to these writers, a task for abstract economic forces but for specific institutions and groups, within the context of the state (which develops citizenship, claims autonomy from other states and the right to use force in a sovereign territory). Attempting to define themselves as independent states with specific cultural and historical inheritances, Latin Americans have carried out the 'huge ideological work to be done daily' (Hall 1991: 26) to determine what they are and are not, and how to imagine a national community. To imagine nations, socio-economic elites have been identified as potential nation-builders, despite their often contradictory identifications. For many elites, Latin American culture and nationhood are lacking (lacking the most desirable Western forms), rather than a solid basis for national pride. In what Yúdice *et al.* call a 'sublation' of centre and periphery, Latin American elites tend to denigrate the local and value the West (Yúdice *et al.* 1992: 4; cf. Hall 1991: 28). Such ambivalent elites are located in contradictory class positions, finding their identities in the interstices of the world's economic centres and peripheries. Although this rightly highlights the easy cosmopolitanism of Latin American elites and their disdain for local cultures, Latin American history reveals that elites have often attempted to direct cultural forms, in order to consolidate their power.

What we are moving towards then is an understanding of the 'imagined community' (Anderson 1991) of the nation which refuses to take on board either a state- or elite-centred notion of the nation-building project, but which on the other hand – as we argue in the following section – does not attribute purity or salvation to the popular cultures of the continent. Rather, the struggles over defining nation and identity through which different groups and institutions (whether cross-class, elite or popular) express their collective subjectivity and political projects, are seen as constitutive – and constituting – the very nature of the national imagined community.

EMBODYING THE NATION

For good reasons, the analysis of national identity and nation-ness has often focused on the state. As a modern regime of power, the state utilizes a series of 'mechanisms of normalization', that come to rest on the body and through which power relations are produced and channelled (Foucault 1977; Chatterjee 1993). Individual subjects are then constituted in and through the relations of state power and the discourses (including nationalism) produced by it. For example, the eugenics debates of republican Latin America during the nineteenth century normalized relations of reproduction and 'race'. The production of national cultures is a huge task, and the state through its institutions and discourses is a major producer. The state generates processes which foster an identification between subjectivity and nation.

Although the modern regime of power has attempted to dismantle and reorganize the identification of subjects within the context of the nation-state, other imagined communities and cross-cutting identities persist. In colonial and post-colonial India, Chatterjee points to the existence of these 'fuzzy' communities in popular political discourses, in which legitimacy and political representation have quite distinct meanings to those attributed them by the state (Chatterjee 1993: 224; cf. Gupta and Ferguson 1992: 17). In the Latin American context, definitions of democracy and participation have distinct (and in themselves diverse) meanings in the many social movements which developed in the region during the 1980s among women, barrio dwellers, indigenous groups, peasants and many others. The social movements represent(ed) not so much the persistence of previous political categories (as in the Indian case), but new ways of doing politics which challenged state-centred political systems (Escobar and Alvarez 1992; Slater 1985; see also Chapter 6 in this book).

This would suggest an approach which emphasizes the multi-dimensionality, agency and self-consciousness of subjects (Reynolds 1994; Cohen 1994), in addition to the practices through which national identity is created and reproduced. There are 'multiple sites in which subject positions [are] produced, and . . . these positions might themselves be contradictory' (Henriquez et al. 1984: 203). In other words, the correspondence of nation and people is constantly overlain by other subjectivities (Bondi 1993: 96; Laclau 1994; Keith and Pile 1993). Overall, the structures of expectations of subaltern and non-state groups with respect to the nation are surprisingly under-researched and under-theorized. National identities are expected to arise from ceremonies and practices which draw citizens into the national sphere. Individuals acquire consciousness of a national identity at the same time as they acquire the national language, an education and other cultural resources. As the nation is 'embodied' in education, secular rituals such as elections, the media and cultural institutions, 'the nation is thus a component in each individual's self- and other-awareness' (Poole 1992: 16; cf. Billig 1995). In order for the nation to become hegemonic in the identities of subjects, elite/official versions of nationalism containing certain histories, images and representations must be shared across class or ethnic lines, in order for an imagined community to be created with 'shared self-awareness'. As nations' content and form are closely tied to specific elites (Balibar 1990), the heroes chosen for state worship and the 'cultic rites' of nationalism vary (Mosse 1975) and cross class lines. Nationalized versions of histories and geographies are widespread in the official institutions with which people come into contact throughout their lives.

Given the class-specific origin of national sentiments, most people may not internalize such narratives; exposure to nationalist discourses does not necessarily entail their adoption (Bowman 1994: 141). Recent studies illustrate the ways in which subjects are either unwilling – or indeed unaware of any need – to take on board any 'national' messages being conveyed. With respect to state rituals in former Soviet countries, Cohen writes,

> these state ceremonies suffered the fate of imposed ritual anywhere: that however well contrived their forms, they could not control the meanings read into them by their audiences.
>
> (Cohen 1994: 163)

Explaining the existence of *internalized* imaginings of the nation alongside the (official) *externalized* national imaginings remains to be done. This is especially true where a minority intelligentsia articulates a national identity on behalf of a (varied and divided) population.[2] In Latin America, 'national groups', such as the bourgeoisies, articulate the imagined community of the nation, yet other classes can be actively involved in these processes for, as Canclini suggests, 'hegemony is allied to subalternity in the practices of power' (Canclini 1993: 34). In this respect, official nationalisms respond to and draw upon popular nationalisms, expanding and rearticulating the images and representations of identity and self-awareness. In other words, the rise of disciplined national subjects has not everywhere seen the demise of alternative senses of community and structures of feeling, which may express anti-state and even anti-national sentiments. Social and spatial relationships and institutions (such as the household, the neighbourhood, the city or a multi-national region) continue to exist within or alongside the nation, and provide potential sites for the creation and re-production of subjectivities. Importantly, these sites are gendered resulting in embodiments which have distinct affiliations to nationhood, whether through representations or practices.

LIVING NATIONS

> partly because of empire, all cultures are involved in one another: none is single and pure, all are hybrid, heterogeneous, extraordinarily differentiated and unmonolithic.
>
> (Said 1994: xxix)

The terms nationalism, the nation and national identities are often used interchangeably (for overviews see Hobsbawm 1990; A. Smith 1991; Anderson 1991; Gellner 1983; Kedourie 1960; Taylor 1989). However, a relatively coherent ideology – nationalism – can be distinguished from the civic and territorial aspect of the nation, and the lived imaginaries of subjects, that is national identities.[3] Emerging as a new term in the nineteenth century (Taylor 1989), nationalism has been further defined as a movement whereby symbols or belief attribute communality of experience to people in a regional or ethnic category (Giddens 1991), or as 'a community mobilized . . . in the pursuit of a collective interest' (Schlesinger 1987: 252; cf. A. Smith 1991: 73).

Both definitions of nationalism place the idea of a 'people' at its centre, yet assume that people are automatically mobilized by such an ideology. However, nationalism is arguably an identity which locates the source of individual identity within a 'people', that is the bearer of sovereignty, central loyalty and a collective solidarity (Greenfield 1992).

National identity can be seen as a wider and more multi-dimensional category than nationalism, as national identity can exist within subjects (collectively or individually) without there being a process of mobilization around a specific goal. National identities can mean different things to different people even within one nation (cf. A. Smith 1991). However, definitions of national identity often refer to the sharing or commonality of a sense of belonging to a specific territory. While national identity is defined by Poole (1992) as 'shared self-awareness', A. Smith (1991: 9) draws attention to the sense of political community and common institutions and rights which reflect and reinforce a feeling of belonging to a bounded territory. For other writers, the commonality which national identity presupposes is based in practice upon processes of exclusion and inclusion, continuously in play (Schlesinger 1987: 260).

Although often represented in official historiographies as a unique form with a unique ancient history, the origin of the nation form is generally associated with the emergence of early mercantile capitalism. Whether first identified in the Americas (Haiti and the United States), or in Europe (with the French Revolution), the current ubiquity and primacy accorded to nations and national belonging are now widely felt to be inevitable (Anderson 1991; A. Smith 1991; Hobsbawm 1990; cf. Balibar 1990; Dofny and Akiwowo 1980). Anti-colonialism may have been key in the origins of some nations, such as in Haiti and the United States, which gave rise to a 'nationalism from below' and distinct forms of class struggle (Blaut 1987; Gomez-Quiñones 1982).

Nineteenth-century international trade and economies provided the context for national identity in the early Latin American republics, which were bound up in 'relations of internationality' (Ree 1992) and specifically in the neo-imperialist interests of Britain (Miller 1993) and other European powers. Some countries flourished in this situation; Argentina was the sixth richest nation in the world by 1900, attracting labour from Italy and elsewhere to its high-wage economy. From the beginning of the twentieth century, the United States had growing economic and political interests in the region, reflected in its economic presence and transformations of cultures. Latin American countries attempted to articulate economically nationalist policies during these years, but as these were coterminous with the United States' influence in the region, attempts were largely undermined. In other words, a long history of neo-imperialism as well as a long history of nationhood make Latin America a unique and uniquely complex region. At economic and cultural levels, colonialism, then capitalism and neo-imperialism drew the region into a specific set of material and discursive relationships through which historical and contemporary identities were elaborated. They have been shaped and intimately linked with the practices and 'structures of feeling and attitude' of (neo-) imperialism (Said 1994: 8). Flows of capital, labour and commodities, as well as the movement of ideas, cultural practices and codes of reception across, into and out of the continent, made the region's historical trajectory an outward-looking and self-conscious one from the start.

Nowadays, while capitalist relations are the most significant in the continent, a variety of forms of incorporation into the market are found. Peasantries whose subsistence depends in part upon non-monetized labour and goods exchange, as well as informal sector workers such as street-vendors and domestic servants, work in economic spheres whose links to capitalist market relations are to varying degrees partial and/or indirect. Resistance to capitalist economics has been identified among certain indigenous Andean groups from the 1850s through to the present-day, while the modern multi-million dollar drug-trafficking business operates at another scale of economic organization.

In addition to experiencing similar economic contexts, Latin American countries shared the colonialists' language (Portuguese in Brazil, Spanish elsewhere), the Catholic religion and a Luso-Hispanic administrative system, all potential foundations for regional and national identities. Declining religiosity and rising secularism of societies are significant seed-beds for modern nations (Deutsch 1953), while the increasing secularism of society makes possible a bounded 'imagined community'. In Latin America during colonial times, the continental spread of Catholicism as an ideology upholding the racial-class-gender status quo, and as an institution, contributed to a region-wide identity especially among the elite. Due to resistance and the uneven conversion of indigenous and rural groups, however, the development of new syncretic religions combining indigenous and Catholic belief was widespread. While the 'continued Catholic substratum' (Brunner 1993: 44) constitutes one dimension of contemporary identity for Latin Americans, the form that this identity takes is highly variable. From the hierarchical, reactionary and anti-secular proposals of the right, to the open, solidaristic and anti-establishment Liberation theologies, the role of religion in identity formation remains a contested field, although Catholicism has long been closely bound up in debates about identity and nation (Lehmann 1991; G. Gutierrez 1992). The contemporary resurgence of religion, especially among various denominations of Protestants – not to mention the syncretic and new religions – is linked perhaps to the difficulty of grounding identity as much as it is a critique of modernity (Beverley and Oviedo 1993: 8–10).

Whatever the attribution of origins, nations and nationalisms are co-terminous with modernity – the rise of the West – and with the relations of capital and neo-imperialism those entailed. Whether in the core metropolitan countries or in the developing periphery, nationhood has been transformed over time as modernity reconfigured social, economic and political relations at local and global levels. In this context, Chatterjee argues that nationalism was presented as the image of the Enlightenment, as a positive, modern and rational form for political development (Chatterjee 1993). Nevertheless, just as the Enlightenment West required its 'Other' for its expression of identity, so too the model of the nation was a contradictory and ambivalent form for those same 'Others' (cf. Bhabha 1990b).

Just as modernity and 'development' as avenues to nationhood have been roundly criticized (Escobar 1988, 1995; Esteva 1987), the early republican dreams of providing national communities through racial mixing or *mestizaje* (especially of Europeans and indians) have been contested by numerous indigenous and black groups throughout the continent. Drawing neither on racial notions nor on Western discourses of progress, popular culture has been seen as an alternative route into national imaginaries and cohesion. Characterized by continual conflict over the meanings and significance of cultural practices and artefacts, popular culture cannot, however, be a straightforward route into a national people. Popular culture is not 'pure' in the sense that it is untouched by global cultural forms and by state appropriation of certain practices – rather it adopts and resynthesizes cultural forms in response to these other arenas of cultural production (Rowe and Schelling 1991). Neither is it closely associated with a specific class group, as its practices and forms may be found in the working-class, peasant or *campesino* class or with elements of the middle-income and elite groups. Given the key role of popular culture in producing subjects, it has been an obvious resource for political and cultural attempts to define nations on the basis of the 'authentic' expression of the people. The state and the media may attempt to selectively reduce and rearticulate forms of popular culture through folklore and mass cultures, yet these (re)appropriations remove cultural process and artefacts from a relatively autonomous popular culture. Popular culture's appropriation and refiguring by the state is found at various times in post-colonial Latin America. Party political rearticulation of populism is one example, as is Peruvian and Mexican *indigenismo* (pro-indigenous measures and culture) in the early twentieth century (e.g. Mallon 1995; Berdichensky 1986). Characterized by the regional elites' 'discovery' of 'pristine', indigenous cultures, *indigenismo* was used in the local bourgeoisies' cultural and political battles to reform the balance of national power; 'part of elite aesthetics tapping into unalienated local cultural capitals' (Yúdice *et al.* 1992: 13).

The cultural purity of indigenous culture can also be expressed in the female form. Paintings of Latin America imagined as a female figure are found in Quito's main public museum and art gallery. In the 'House of Culture', one oil painting entitled *América* shows an indigenous woman in the foreground with indigenous children playing, while Spanish soldiers with helmets, shields and swords upheld approach from a ship in the background. Above the woman sits an ethnically white woman with a crown and sceptre, floating on a cloud and resting against a lion. The indigenous 'America' wears a loose white tunic, a coloured cape and a feather headdress, and is carrying a bow and arrow.

Popular culture, while often taken as the simple expression of 'the people', can also more usefully be seen as the diverse discursive sites where popular subjects are formed and reproduced. To take one example, the use of *traje* ('indigenous clothing') in Guatemala has become a symbol of national identity for the state, and is purveyed in international beauty contests, tourist brochures and business literature as the expression of a true national essence, shared by the indigenous and *mestizo* populations of the country. The indigenous groups in the country, from whom the *traje* was appropriated, continue to use the clothing within their own homes in addition to local languages, customs and rituals, as the basis for expressing their own identities, which are critical of the state and its use of their clothing (Hendrikson 1991). Jameson's suggestion that such expressions are national allegories is thus widely off the mark (Jameson 1984; cf. Yúdice *et al.* 1992: 17).

Contemporary nations are often a mixture of popular, official and creole expressions of nationalism (Anderson 1991; also Taylor 1989), and depend upon the emergence of a commonly shared public-national sphere in which symbols, practices and institutions elicit a widely shared comprehension and identification. It is in this sense that Laclau and Mouffe identify the nation as 'at the same time a fiction and a principle organizing actual social relations' (Laclau and Mouffe 1985: 119). The task of articulating a national fiction and explaining national social relations is often given to – or appropriated by – intellectuals, as secular exponents of a national vision (cf. A. Smith 1991; Gellner 1983). Nationalist histories are constructed by intellectuals around the 'rediscovery' of events, their reconstruction in a new historical narrative and imagined geographies, and may entail fabrication and insertion into invented ceremonies (Giddens 1990; Hobsbawm and Ranger 1983). To ground national history, events are linked, often by intellectuals, with certain landscapes, sacred sites, historical monuments and places in the nation (A. Smith 1986; also Daniels 1993; Schama 1995). In Latin America, intellectuals have played a role in defining national identity (Lomnitz-Adler 1992; Brading 1991). However, in a region where self-representations and identity have been so shaped by colonial and neo-colonial representations, national intellectuals have been only partially successful creators of national identities. Latin American state-sponsored historiographies created heroes (such as Juan Santamaría in Costa Rica) who could represent the nation by suppressing elements which cut across national unity; heroes, 'stripped of class representation, regional affinities or religious zeal' (Rowe and Schelling 1991: 229).

On what basis then does a notion of 'the people/*el pueblo*' provide for the emergence of a nation? Rather than see popular culture as a metaphor or an abstraction of the 'national', the emphasis in this book is on the cultural practices and the 'local codes of reception' which characterize popular practice and discourse in their quotidian definitions of nationhood and affiliations to place (Brunner 1993). Central to these definitions are the constant struggles over interpretive power (Canclini 1993). An emphasis on cultural practices and codes of reception allows for understanding the selective resistances and adoptions within specific contexts have, at several points in Latin American history, been shaped by the use of violence (Rowe and Schelling 1991: 8). Yet despite suffering violence and transformation, popular cultures offer discourses and practices generating alternative ideas about nation and community.

Globalization has meant the emergence of often dynamic and innovative hybrid cultural forms and elements for identity construction (Yúdice *et al.* 1992; Canclini 1993). Such hybridization of cultural forms and bases for identity – called cultural reconversion in Latin America – is widespread.

Western film characters for example get culturally reconverted into low-income heroes with local affiliations. One example is Superbloque, modelled on the cinematic Superman, who has become a Mexican popular hero fighting on behalf of neighbourhoods threatened with eviction or neglect by the state (Figure 1.1). Highlighting the ways in which local practices mediate global cultural circuits and synthesize new hybrid forms, a focus on cultural reconversion is not, however, to value the 'local' over the 'global', as if they were discrete categories with systematic features; identity is not simply 'Made in Peru' or 'Made in Brazil', but neither is it 'Made in USA'.

Stuart Hall, in discussing identities in the globalizing world, argues that

> the recreation, the reconstruction of imaginary, knowable places in the face of the global post-modernism which has, as it were, destroyed the identities of specific places, absorbed them into this post-modern flux of diversity.
>
> (Hall 1991: 35)

Separating out (multiple) identities from affiliations with place remains problematic, risking the attribution of autonomy in social identity formation (cf. Hall 1991: 36). Such problems arise in part because spatial terms are used without clarifying whether they are used metaphorically or literally (Gupta and Ferguson 1992; Watts 1992). Latin American writers often return to this issue and the question of the nation's relationship with the local and the global, the nation being the focus of discussions around citizenship, democracy and diverse political projects (Abán Gómez *et al.* 1993; CAAP 1981; Díaz-Polanco 1989). The national spatial level, it is often argued, is *the* level at which local codes of reception and cultural practices are constituted as the medium for identity construction. However, with no pre-constituted subjects, identity formation is not purely and simply a reflection of a place in the world where subjects are located. Rather, the relationship between pre-constituted subjects and the local/national/global is one which has to be articulated and rearticulated, expressed and received, made and remade.

PLACING THE NATION

> If agents were to have an always already defined location in the social structure, the problem of their identity would not arise.
>
> (Laclau 1994: 2)

In the potential overlap of territory, culture and population of the nation, there can be a disjuncture between the national place and national identity; the 'space of the nation' can be imagined by populations which have no 'place' in which to express and consolidate that identity. Such disjunctures, or 'dis-located senses of place' (Jacobs 1995; Bowman 1994; Malkki 1992) draw attention to the work done by imagination and discursive formations in (re-)creating linkages between people and place. Even where populations and 'their' place are together, identities are expressed through 'imaginative geographies – the term is Said's (1978) – by which the differences and

Figure 1.1 Superbloque, a character culturally reconverted from Superman

distinctions between 'Us' and 'Them', and between 'our place' and 'their place' are discursively imagined and articulated. Such imagined geographies can provide the basis for a shared identity, articulated through a sense of sameness in social features and a sense of shared space/place, a 'homeland'. In many contexts, the national icons of territory, language and culture are assumed to overlap and frame each other. However, a direct relationship between place and identity has become increasingly untenable – if it ever was (Hall 1991; Bhabha 1990a; 1990b; Keith and Pile 1993) – in a world of numerous diasporas, and of increasing global interconnectedness (Bird et al. 1993), and draws into sharp relief the creative and non-essential nature of the linkages between people and nation. In these 'postmodern geographies' (Soja 1989), the relationships between place, nation and identities are increasingly being interrogated as to their permanence and social relevance, in a world where nationality has only a partial role to play in individuals' participation in the global economy and the power relations it entails.

The contexts in which identities are instantiated, selectively reproduced and utilized, represent a 'multidimensional set of radically discontinous realities' (Jameson 1986: 349) through which multiple subject positions can begin to emerge and coalesce in individuals. Such a formulation of discontinuity in place and identity moves away from a naturalization of the place and identity relation. Where identity reflects the place of subjects, and the place is defined by the subjects in it, a naturalization occurs, risking stasis and closure (Keith and Pile 1993). Post-coloniality provides for the constant movement and juxtaposition of identities and place (Loomba 1993; McClintock 1993b), although nationalisms are not necessarily redundant in this context. Places (re-)emerge as the possible (contradictory) sites for the creation of identities. Functioning as 'symbolic anchors' of identity (Gupta and Ferguson 1992: 11), places (re-configured by late capitalism, migration and media) can call up and reinforce a 'spatial commitment' (Gupta 1992: 63) on behalf of populations whose identity and sense of community can be expressed in terms of that configuration. As Watts points out, it is a question of seeing

> how individuals are interpellated by multiple and often contradictory cultural and symbolic practices, rooted in historically constituted, yet increasingly global, sites.
>
> (Watts 1992: 124)

Given the mutual imbrication of social and spatial practices, the existence of power relations at the heart of societies must be considered (Massey 1993; Rouse et al. 1992). Territory–population relation is naturalized through the power-filled actions of carving out of a (unique, national) place from (usually varied) material environments, and the creation of that place as incomparable, a unique place for the expression of identity (cf. C. Williams and Smith 1983: 503). The 'spatial discontinuity' of the nation (Fox 1990) is simultaneously both the grounding of and the expression of the national people as different, an expression of power and will to order. The creation of a social space in which to express belonging and at the same time to define those which are 'Other' is the effect and practice of the socio-spatial power relations of the nation. And if the creation of a past for a national community and the expression of identity has been commonly recognized (Renan 1990; Friedman 1992), the new concepts of space-time imply a reconceptualization and a 'remapping' of the nation and its identities. The boundaries between nations reinforce territorial segmentation at the same time as they reinforce notions of purity and sameness within the territory, and difference and impurity outside the territory (cf. Bhabha 1990a). In other words, while the nation-space may exist discursively as the overlapping of territory, culture and population, it is also very much grounded in practices and materiality, and both narrative and material dimensions need to be conceptualized simultaneously. Everyday practices produce and reproduce national identities in a variety of sites, from the home and neighbourhood to the workplace and public sphere. Rather than being merely narrations (Bhabha 1990b), national identities and nations are embedded in the material and imaginative spatialities of collective and individual subjects.

Our emphasis on the materiality of identities is prompted in part by the discussion in Laclau and Zac (1994) on the disappearance of political opponents under certain Latin American repressive military regimes. While the military recognized the existence of the *desaparecidos*, the disappeared, they denied responsibility for their disappearance. Laclau and Zac suggest that the disappeared are a 'space of suspension, which is both part of, and excluded from, the realm of "society"' (1994: 34). To our mind, the *desaparecidos* are a prime example of the field of power operating in the military state through a series of substantive *material* exclusions to define 'the people' of the nation. In this case the authoritarian nation-state excluded political activists and anyone challenging their legitimacy. More than being an example of the state attempting to 'culture thought' among citizens (Cohen 1994: 162), the Latin American states acted as an apparatus of discipline and productive power intervening materially and violently in a variety of sites to reinforce their idea of the nation (cf. Perelli 1994; Fisher 1993).

With globalization in the late twentieth century, the context for national ideologies and national identities is being transformed, as labour migration, capital mobility and the creation of mass cultures change national imaginings. While globalization is a highly uneven process in Latin America, the continent is deeply implicated in increasing global integration and resulting transformations. In the political and economic spheres, globalization has entailed the privatization of previously nationalized companies, the dismantling of already-weak welfare states, and rising technocratic control of economic and political decision-making (Yúdice *et al.* 1992: ix; Lehmann 1991; Larrain 1991). Privatization of what were historically 'national' cultural spheres and institutions – such as museums, galleries, and support for the arts – has occurred, alongside the mixing of high and popular cultures.

Conflicts arising from subordination in the global market and from 'segmentation and segmented participation' in the global circulation of symbols are all implicated in the critique of the nature of Latin America's insertion into modernity (Brunner 1993: 41; Beverley and Oviedo 1993). In this sense then, shifting and globalizing economies and cultures are not characterized by a resolution of conflicts but their transformation, in which the 'nation' holds out an ambiguous promise of community and continuity (Albó 1993; Yúdice *et al.* 1992; also Rowe and Schelling 1991). The rise of globalization, the emergence of plural voices and the spread of mass cultures all prompt Latin Americans to reconsider their national identities and national 'imagined communities' in the contemporary world. Early republican leaders had faith that certain national imaginings would mobilize communities. Yet these narratives operate distinctly under globalization, and nationalist projects of attempting to fix relationships between people and place now take place within globalizing transformations. At the same time that attempts are made to create territorial identities of nationhood (cf. A. Smith 1991), the mingling of two or more localities is occurring (Appadurai 1990; Bird *et al.* 1993). Consequently, 'national differences are reconfigured through transnational interaction' (Canclini 1993: 34, 39), however bounded legally or constitutionally societies may be.

DE-CENTRED NATIONS

In this sense, social subjects, institutions and communications interact across a variety of different spatial scales, such that earlier formulations of 'concentric circles of allegiance' (C. Williams and Smith 1983: 515) presented too simplistic a picture of overlapping identities. Interactions and interrelationships at numerous spatial scales define nations and nationhood as much as the territorial boundaries and official maps of the state, based on the sovereignty principle. Consequently, subject positions such as citizenship

> ought to be theorized as one of the multiple subject positions occupied by people as members of diversely spatialized, partially overlapping or non-overlapping collectivities.
>
> (Gupta 1992: 73)

The diversity and multiplicity foregrounded in the work of Gupta and Chatterjee, among other writers, are based on the understandings from poststructuralism that identities are not static and unitary but shifting, contingent and lacking in fixity. This malleability however is not completely free-floating but relates to conceptions of time and space, and the relationships between histories, cultures and biographies. It is these articulations which, following Foucault, have come to be understood as 'sites', an understanding which has framed our conceptualization of nations, nationalisms and national identities in Latin America, and more specifically Ecuador. To work with the concept of 'sites' implies distance from earlier accounts of the social, which saw society as an organic whole or as a system with unity that can be uncovered by the application of sociological (including Marxist) knowledges. Instead, society is seen to be itself fractured and constantly in the process of constitution and re-constitution in which specific sites play key roles. These processes and sites are permeated by power relations which do not necessarily have a single dynamic which, once uncovered, would unlock all other forms and expressions of power. As Seidman notes,

> [Foucault] imagined modern societies as fractured, lacking a
> center that gives to them a unity and telos. Neither the state ı
> economy is the social center; no one . . . conflict, not class conf
> gender, sexual, ethnic or religious conflict, carries any obviou. . .
> or political primacy.
>
> (Seidman 1994: 225)

Clearly, this multiplicity mirrors the multiplicities of subject positions and identities and provides the basis for the de-essentializing and de-centring of the social. It is this understanding of the social as de-centred that is one major element in what Rattansi (1994) has called the 'Postmodern' Frame. It does not imply a retreat into Baudrillard excesses nor does it express a view that 'there is no society'. On the contrary, it speaks to and from within the myriad societies and the complexities of social formations and the sociological enterprise which seeks to address the formation, re-formation and de-formation of the social. In this imaginative leap from the structural to the post-structural, an emphasis is placed on the suffusion of power organized in disciplinary modes. One of these modes is the nation, discursively organized in a variety of sites, meeting opposition and counterclaims that seek constantly to disrupt the process. These disruptions are a crucial part of current nationalisms in Latin America, exemplified by the indigenous movements. However, as mentioned above, this should not be understood as a binary between the official version of nationalism – THE national project – and an alternative, the popular. It is clear from Derrida's work that binary oppositions are the major organizing principle for modern knowledges constructing identities in opposition and fixity, rather than as contingent and provisional (Derrida 1977, 1978).

The work of Laclau and Mouffe (1985) has been one major attempt to use a de-centred notion of the social in relation to an analysis of political trajectories and, as earlier sections suggest, it has major implications for the study of national identities as a political terrain. The vision of politics within Laclau's work is of a form of 'trench war' in which positionality, collectivities and outcomes cannot be read off from the predetermined positions, but are constantly being made and remade within the shifting terrain of power relations. As the social is 'always subject to subversion' (Rattansi 1994: 31),

> the social is a 'decentred structure' composed of practices of centering,
> the construction of power centres around the nodal point of articulation.
>
> (Rattansi 1994: 31)

It is precisely this attempt at centring which is at issue for national projects and national identities. Our work has attempted to elaborate and analyse the ways in which centring and de-centring occur simultaneously within the realm of the nation. In part, this returns us to issues of time and space, the writing of histories and the drawing of geographical and social boundaries as spaces for the imagination. What is all too evident is the fragility of these attempts at centring and the strength of the fractures. These 'centring' projects – national, racial, economic – structure in myriad ways (some of them explored below) the incorporation, discipline and 'domestication' against which the resistance, destabilizing effects of alternative discourses and practices work. Sometimes, resistance is articulated through the collective subjects, while at others it is more muted and diffuse. In nationhood and national identities, a grand narrative of the nation is organized around representations of place and people through history, with its own seductions, the most basic often being 'we are not they'.

In such circumstances, what possibility exists for the emergence of an imagined community with shared self-awareness, if the forms by which national imaginings are conveyed and mediated are so differentiated? As discussed above, the emergence of global economic and information networks does not necessarily interfere with the 'local codes of reception' operating in Latin American social and cultural practices. Nevertheless, the extent of 'local' codes of reception has not been documented or theorized. In this respect, a question remains about the extent to which different groups within 'national society' share these 'local'/'national' codes of reception. It is precisely the gaps between these representations of the nation that recent work has begun to explore. The existence of *alternative*, that is non-official, accounts of nationhood and national identity have been recognized in post-colonial and post-modern societies around the world (Loomba 1993: 306). In the continuous redefinition of the nation by groups, 'rival versions of national identity are produced, on a terrain of contestation' (Schlesinger 1987: 257). It is on this terrain of contestation that Bhabha identifies an 'ambivalence' in the definition of national society. This ambivalence arises as the groups defining themselves and others as 'the people' – the constituent subjects of the nation – do so not only on the basis of affiliation, but also crucially by disavowal, displacement, exclusion and contestation (Bhabha 1990b: 5; also Schlesinger 1987: 260).

The power relations behind such exclusions go against an interpretation of nationality in terms of complementarity or efficiency of communication among national subjects (cf. Deutsch 1953). Such formulations deny subjects' unequal communicative ability and access to discursive power in the nation. 'Imagined communities' have to specify *who* imagines the community and *how* they do so, and what differences in communities consequently arise. The 'disputed field' (Fox 1990: 7) of the nation is subject to 'vacillating representations' (Bhabha 1990a: 297), none of which are fixed but which are plastic and contingent, depending upon power relations and the creation of discursive fields. Especially perhaps in multi-ethnic, post-colonial metro-politan and developing countries, the 'reading' of the imagined community from the discursive and representational resources offered to subjects is inevitably problematic. In other words,

> while we know that national identities are imagined communities . . . , we know little about how ideas of nation and identity are produced and reproduced as elements of political struggle in which other possible dimensions of identity formation – race, gender, class and locality – are likely to become conflated with ethnicity and territoriality.
>
> (Reynolds 1994: 243)

The creation and embodiment of national identities is thus problematized at the point of creation – what shape to give the imagined nation to capture the imagination of its multifarious subjects? – as well as at the point of reception – how will the diverse local codes of reception *within* such nations respond to the image on offer?

REMAKING THE NATION: NEW APPROACHES TO NATIONAL IDENTITIES

In order to provide a new vocabulary to describe and analyse relationships between identity, place and subjectivities, we use two new concepts, first, *geographies of identity* and second, *correlative imaginaries*. Geographies of identity (see Chapter 5) can be defined as the senses of belonging and subjectivities which are constituted in (and which in turn act to constitute) different spaces and social sites. Geographies of identity become embodied at the personal level, deriving from individual subject's biographies and interpretative schema, but the focus in this book is on *collective* or shared geographies of identities, the 'local codes of reception' and imaginative geographies which ground collective identities. The collective subjects constituted in relation to these fluid senses of space are conceptualized as unfixed subjects, subjects which defy simple binaries (such as public/private, national/anti-national). Geographies of identities are lived geographies – the latter being what Lefebvre (1991) calls the spaces of representation – through which identities are continuously constituted and undone by the daily social relations, technical materialities and discursive forms they are lived through. Geographies of identities are also constituted in the spheres of the imagination and the representational, what Lefebvre calls the representation of spaces. The idea of geographies of identities draws from Said's imaginative geographies a sense that structures of belonging, difference and spatial organization are constantly drawn up and peopled influencing the constitution of identity, in relation to an Other.[4] Given the mutual embeddedness of sociality and spatiality, the geographies of identity are also conceived as racialized, gendered and class-based.

The racialization of geographies of identities occurs through the attribution of racialized groups to particular areas (of the nation, the city and so on), in the collective imaginations of citizens. To give an example, the geographical imagination of Peruvian citizens and the state in the post-colonial period has associated the indigenous population with the Andean mountains, and has seen 'development' as a task to overcome the physical and social 'obstacles' represented by this racialized geography (Orlove 1993). In Ecuador, our research found that 'popular expressions' of identity racialized the geography of the nation. Afro-Ecuadorean populations mapped themselves out on to the national space to cover a wider area than imagined by other race-ethnic groups (who either 'forgot' to map them on to the country or imagined them in small, self-contained areas). In Ecuador, the recent emergence of alternative accounts of national identity provides insights into the dynamics of an ambivalent 'nation', where the apparent fixity of the official nation is constantly undermined by the expression of alternative accounts of nationhood.

The seductions of nationalism, and national identities are, in part, an affair of the heart – what Anderson (1991) has called a 'political love' – which for all its romance and emotion is still bound to the ideological work of nation-building. The term does, however, capture the huge emotional investments that people make in relation to their national identities, which states can call upon in times of crisis and on which politicians lay claims. We have sought to re-examine this issue, through the concept of *correlative imaginaries*, which is more consistent with the account of a de-centred social and distinct from the fixity implied by Williams and Smith's (1983) 'concentric circles of allegiance' (see Chapter 4). Correlative imaginaries generate and sustain an ideational horizontal integration with a shared space, through a form of interpellation which correlates subjectivities and social spaces. This allows individuals to place themselves within a 'frame' and to produce a form of doubling between the self and the social, within specific sites. This placing – the generation of positionality – can, therefore, be said to be part of the ensemble of relations that produces national identities. Although there are many sites in which correlative imaginaries can be produced, we concentrate upon their genera-tion within forms of popular culture, especially televisual forms including football and *telenovelas* (television soap operas), as well as newly emerging popular religions, all of which have great emotional appeals. The imaginary of the nation is usually bound to a 'fictive ethnicity', organized around an homogenizing account of 'race' and nation. In Chapter 2 we examine this, deconstructing this version of the centred nation.

2

'RACE', STATE AND NATION

Clara is 40, a woman of African descent who works as an *empleada* (a maid or domestic servant) and who tries to live in a room no larger than a cupboard with her 11-year-old son in an expensive district of Quito, Ecuador. They share a bed but mostly Clara sleeps little, working from dawn until dusk for minimal wages, little holiday and without pension or insurance protection. It is not simply fortuitous that Clara is black; domestic servants are of African or indigenous descent throughout Latin America. Companies selling washing powders know this well and advertising is aimed at black women. But, and it is a very large but, the racialization and feminization of domestic work is so much part of the commonsense reality of Latin America that it is 'invisible', hidden by a series of articulations that appear initially to be contradictory. The first of these is that people of African descent are highly visible and subject to a series of disciplinary discourses relating to culture, history and colour. These discourses contradict the official ideology of 'racial democracy' articulated throughout the continent but most highly developed within Brazil. Similarly, but bound to conceptions of 'otherness', the visibility of people of indigenous descent is no less disciplinary but is articulated with conceptions of culture rather than colour. Thus, there are a number of racisms in play but domestication and feminization make difference and visibility perform a vanishing trick and, within their working lives as *empleadas*, women of indigenous and African descent are made invisible. The articulations between gender and the sphere of the home, constructed as a private sphere, provide a site within which both domestic employment and its racialization are made invisible (Radcliffe 1990b). There is no surprise here, as Chapter 6 elaborates: ideologies of domesticity and the importance of women as wives and mothers are the dominant discourses around femininity. The labour of domestic relations is written out of public discourses because, of course, the wives and mothers who are celebrated are the very wives and mothers who do not perform this labour but hire domestic servants who are exploited, dehumanized and denied a home life of their own in the process. Clara's story is not, therefore, an aside or a victim's tale: it holds within it the complex ensemble of 'race', state and nation, that is, racialized and feminized labour within the context of citizenship, the state and the home and their articulation with the nation, issues to be explored in this chapter.

RACISMS

The theorization of racism is an arena of contestation in which earlier views of racism as individual prejudice have been subjected to critique (Rattansi 1992) and found wanting, while the equally powerful view that racism is tied to class has been re-examined in the light of an understanding of racism as more complex, multiple and shifting. Anderson (1991) in a memorable phrase does suggest that racism 'dreams of eternal contaminations' while 'nationalism thinks in terms of historical destinies'. But, he concludes, 'The dreams of racism actually have their origin in ideologies of class, rather than those of nation: above all in claims to divinity in rulers and to "blue" or "white" blood and "breeding" among aristocrats' (1991: 149). While notions of breeding, blood and purity have certainly marked racist discourses these are often woven together via a national or, in addition to, a class story. Like Raymond Williams (1987), who views the nation-state as the organized and orderly market for international capital, the issues of capital and racism are articulated for Anderson in the colonial 'adventure' which offered a model of European racism which was exported around the world. Yet, the discourses of European racism were considerably more complex including, in relation to Spanish Conquest, a well-developed debate from the sixteenth century among Jesuits on the nature of souls and the admissibility of souls among the indigenous populations of South America.

It is not surprising, therefore, that Homi Bhabha, for example, should take issue with both Anderson and Foucault. Bhabha (1994: 248–9) suggests that their views constitute a conjuring trick in which racism is simultaneously tied to modernity and the rise of the West but as an archaic hangover bound to an ahistorical universalism. Bhabha writes:

> In placing the representations of race 'outside' modernity in the space of historical retroversion, Foucault reinforces his 'correlative spacing'; by relegating the social fantasy of racism to an archaic daydream, Anderson further universalizes his homogenous empty time of the 'modern' social imaginary.
>
> (Bhabha 1994: 249)

These accounts are, for Bhabha, a form of myth making in which the notion of the progressive values of modernity are maintained and the contents of racism are written out, in this way contributing more generally to the homogenizing national story which thus 'deprives minorities of those marginal, liminal spaces from which they can intervene in the unifying and totalizing *myths* of national culture' (Bhabha 1994: 249). It is precisely this issue that is central to this book – the impact that racism has on de-centring the nation – and the ways in which minorities in those 'liminal spaces' have been able to reconstruct the national story and insist on the plurality that was constitutive of the modern project.

As an alternative Bhabha suggests

> that we see 'racism' not simply as a hangover from archaic conceptions of the aristocracy, but as part of the historical traditions of civic and

liberal humanism that create the ideological matrices of national aspiration, together with their concepts of 'a people' and its imagined community. Such a privileging of ambivalence in the social imaginaries of nation*ness*, and its forms of collective affiliation, would enable us to understand the coeval, often *incommensurable* tension between the influence of traditional 'ethnicist' identifications that coexist with contemporary secular, modernizing aspirations.

(Bhabha 1994: 250)

Such a view of racism has a special resonance for Latin America where the ambivalence and tension have woven together a narrative organized around 'ethnicist identification' which has been produced precisely to negate ethnicity and to create, indeed, Balibar's (1990) notion of a 'fictive ethnicity', intimately linked to the rise of the nation-state and the power of Conquest. In the language of 'racial democracy' and *mestizaje*, hybridity is the norm and exclusive ethnicities are archaic, while the modern project within Latin America is tied to *mestizo* identity, but the language of whiteness remains as expressed by General Lara in the now famous phrase:

Todos nos hacemos blancos cuando aceptamos los retos de la cultura nacional (We all become white when we accept the goals of national culture).

(quoted in Whitten 1981)

It is possible within these official, public discourses to nullify and render absent the issue of racism, but as the responses to General Lara's words in our interviews in Ecuador suggest, the issue is ever present. From the rural indigenous woman who responded with the fear that 'they want to kill us all' to the African-Ecuadorean man who said, 'We are black. We can never be mestizo and part of the national culture. They will never allow it'. He also said, 'racism and our economic predicament' were shared across Latin America. Manuel was not alone in his view; it was repeated by many African Ecuadoreans, one of whom, Nina, said in response to General Lara's words, 'No way, no way I would compromise my race, never, not for anything in the world.' Quite clearly, there are powerful counters to the hegemony of notions of racial democracy and *mestizaje* and these contestations draw on an historical perspective that is both local in the ethnicization of locality and global in relation to diaspora and migrations.

It is important, however, not to privilege the issue of racial democracy (to which we shall return) without attention to the complexity of ethnicities in Latin America. The West may well have been surprised to learn of the election of Fujimori, a man of Japanese descent, to the presidency in Peru, ignorant of the size of the Japanese 'community' in Peru and shocked by the bombing in Buenos Aires of the Jewish Community Centre in 1994. Argentina has the largest settled community of Jews, numbering 250,000. The place of the Jews in the national story of Argentina is a complex and contradictory one with the processes of racialization, anti-semitism and otherization oscillating with forms of economic and political integration (see e.g. Guy 1991; Metz 1992). Equally, throughout Latin America there are communities of Lebanese and Palestinians alongside the European migrations. This complexity is born of the myriad diasporas marking a global history which has given Latin America a specific degree of hybridity and complex inter-ethnic relations. Such invisibility is officially sanctioned in the common practice of not collecting ethnic statistics in the census data for states; Ecuador and Peru, for example, do not have official statistics on ethnic composition. This is an important part of the ideologies of racial democracy discussed below.

Historically, Conquest brought an encounter between Europe and the Americas and was a crucially formative period for the generation of European identity. As Stuart Hall suggests:

> The story of European identity is often told as if it had no exterior. But this tells us more about how cultural identities are constructed – as 'imagined communities', through the marking of difference with others – than it does about the actual relations of unequal exchange and uneven development through which a common European identity was forged.
>
> (Hall 1991: 18)

In the 1990s too, the complex interweaving of Judaeo-Christian and Islamic cultures, so well defined in Spain historically, has continued to reassert itself within the fractured identity that is European, precisely because the homogenizing Christian, white account of European identity has *always* been inherently unstable.

However, within this imaginary of the 'civilizing Christian mission', the Conquest took place alongside the dreams and visions of avaricious men in search of Eldorado.

Just as European identities and colonialisms were not unitary, so the ways in which racisms were generated and sustained cannot be thought in a unitary way. From the historical accounts of the colonial periods it was clear that there was in play, from the earliest times, a bifurcation in the ways in which the processes of racialization were effected (Galeano 1973; Williamson 1992). Thus, the indigenous peoples were constructed as truly the 'others' of the Americas in relation to culture as well as skin colour, while the enslavement of Africans brought black people to the continent as labour power, as chattels. In relation to African people, discourses around cultural difference were subsumed in the pseudo-scientific version of racism which privileged the body and its functions. These differences have been explored in the most telling and dramatic way by Michael Taussig (1987) in his account of the rubber (robber) barons of Colombia and Ecuador in the nineteenth century. Although it might have been 'rational' within the conditions of capitalist exploitation to emphasize economic motivations, the owners of the rubber plantations enacted genocidal regimes upon indigenous workers. At the same time, imported African labour was used to control indigenous labour, setting in place a contradiction between African and indigenous descent peoples that survives today in the *haciendas* of the Ecuadorean Sierra where the current owners employ black Ecuadoreans as private security forces. This example shows how the processes of racialization can be used simultaneously to marginalize and co-opt different ethnicities, demonstrating the dynamic and contingent character of racisms.

The violence of colonial history is shown in myriad ways across Latin America. It is a history of genocide and resistance by indigenous peoples and attempts by the African population to generate a freedom song. Slave revolts marked the early period; one of the most important in Choca, Colombia, took place in 1728 and while it was put down by Spanish fire power, the strength of the revolt had a marked impact on the gold mines, a key sector in the regional colonial economy (Sharp 1976). The sixteenth-century kingdom of Esmeraldas was run by the black population (in what became Ecuador), another example of the ways in which the African populations sought to control their own destinies in the region. Although Peru became independent in 1821, the struggle to free the 50,000 enslaved African Peruvians took another thirty years (1855). The liberators met with organized resistance to the abolition of slavery from the slave owners, who in turn were well served by British travellers endorsing the view that Peruvian slavery was more humane than elsewhere. Such a benign construction has been dismantled by Blanchard, who states that 'Peru's slaves were an abused and exploited sector of the population' (1992: 87). Equally, Chinese labour and European migrants 'were exploited and mistreated' (Blanchard 1992: 142). Abolition did not end racism, as illustrated in an article in a Cuzco newspaper in 1855 which 'accused blacks of being "the worst race of the human species for their intellectual unsuitability" and "a terrible plague for Peru"' (Blanchard 1992: 218). These views were endorsed by other writers and the suggestion made that the black population should be annexed to a colony in the Amazon region, a classic coming together of 'race' and place in a 'geography of exclusion' (Sibley 1995).

'RACIAL DEMOCRACY'

In Brazil, as both Hanchard (1994) and Reid Andrews (1992) detail, black protest has been a crucial element in the history of the Brazilian nation-state, as a response to the fact that 'Portuguese state policy made black slavery the foundation of Brazil's social and economic order during three centuries of colonial rule' (Reid Andrews 1992: 148). Consequently, Brazil was the last republic to abolish slavery in 1888 and the first to articulate conceptions of 'racial democracy'. As Hanchard (1994: 74) notes, 'the ideology of racial democracy has become national commonsense that has informed both popular folklore and social scientific research.'

Central to this 'folklore' has been the person and position of Pelé, the great black footballer (Simpson 1993). Pelé's representation has allowed politicians and some people in Latin America to disclaim racism as a European problem. But 'racial democracy' is, in effect, the process of *blanqueamiento* or 'whitening' through miscegenation and, as Wade comments, the emergence of a large mixed intermediate group . . . has established the myth of Latin American 'racial democracy' based on the predominance of the *mestizo* and the *mulatto* and in which racial marks are no barriers to marriage and mobility.

(Wade 1986: 16).

What this hides is not only routine discrimination in relation to resources, suggests Wade, but also the 'attempt to escape blackness'. As we have suggested, this escape is marked by violence, a violence of disavowal and rupture with historical and cultural roots (Westwood and Radcliffe 1993). The disavowal was part of an organized 'racial project' which emphasized 'racial democracy' and whitening in the aftermath of the abolition of slavery. Brazil's 'racial project' was elaborated in relation to capitalist development and the markers of modernity. Thus, Hanchard notes,

the racial composition of Brazil brought consternation to many Brazilian elites. . . . Comtean positivists and eugenicists, partners in social alchemy, confronted Brazilian elites with racialist formulations of its own. If Brazil was to be modern, with the non-white hue of its people, it could not possibly be like Europe.

(Hanchard 1994: 53)

In order to sustain a notion of development and in organizing the racism of the nation-state, Brazil passed an immigration law in 1890 excluding Africans and Asians from entering the country (it was enforced until 1902). The racism of the legislation was consistent with the emphasis upon 'whitening' Brazil, especially in the civil service and political class more generally. Later, it was Freyre's work (*The Masters and the Slaves*, initially published in 1933) that sought most comprehensively to articulate an account of Brazil founded upon a dual heritage, European and African which could generate a very specific 'race' of people. As Hanchard notes,

Freyre's vision should be seen not merely as one about race and racial difference, but as a subset of a national project of conservative liberalism, complete with the paternalism and patron–client relations that have marked Brazilian society and culture from colonial times to the present.

(Hanchard 1994: 54)

Hanchard's analysis attributes an important role to Brazilian intellectuals in the generation and sustenance of a racial hegemony located in the myths of racial democracy, like the claim that racism is a European or North American problem and the naturalization of the power of whiteness. Racial projects are to be found throughout Latin America that seek to privilege a 'fictive ethnicity' woven together within the discourses of racial democracy. In Mexico, for example, as Hernandez-Díaz (1994: 78) comments, 'The Mexican *mestizo* identity is an ideology that has been implemented as state policy, and this policy has statistically transformed Mexico's demographic composition'. Wade in his work on Colombia, like Hanchard (1994), emphasizes the important symbolic power of *blanqueamiento*, which as a process is tied to economic and political relations but 'is also a dynamic that involves culture, identity and values' (Wade 1994: 341) and cannot be reduced to the economic and political moments. Thus, Wade's discussion of whitening weaves together the complexities of strategies relating to the valorization of whiteness. It is a racial and colour hierarchy founded upon the negative evaluation of blackness. What this produces, suggests Wade, are a series of subtle and often contested manoeuvres between, within and across the colour hierarchy. White and black, in the variety of shades to which these homogenizing notions refer, intermarry and have relationships. Moreover, the discourses which surround these processes are highly gendered. Black men with white women suggests the power of black men to enter the world of whiteness via sexual licence. However, black women in relationships with white men are differently assessed to the point where their actions are regarded as a betrayal of blackness and 'the black race' (Wade 1994: 295–314). Clearly, this mirrors gendered relations of subordination and dominance in an idealized discourse; the lived experience of power relations between black women and men, black and white, may be very different. Equally, while the notion of racial democracy should consistently offer greater status to the *mulatto/a* (person of mixed white and black 'race') population, this is not necessarily the case. A hybrid population may disavow the symbolic capital of miscegenation in ways that value such capital as a deficit, no capital at all. The literature on racial discourses include an elaboration of the negative aspects of hybridity drawn from eugenics (Young 1995), which is reworked within discourses on authenticity and betrayal from within the white and black worlds. Thus, Gilroy (1993: 7) notes, 'sexuality and gender identity are the other privileged media that express the evasive but highly prized quality of racial authenticity'. These considerations suggest just how pervasive and pernicious the ideology of whitening actually is in the Latin American context and how far it exposes the myth of 'racial democracy'. The symbolic is, however, in play in other ways too, in relation to migration to the urban world, and its modernizing and valorizing power, which confers a degree of mobility and thereby distance from a racialized peripheral status even, suggests Wade in the Colombian context, for black working-class subjects. Whitten (1981) writing in an earlier period of capitalist development in San Lorenzo, Ecuador, suggests:

> The mass of people identifying with the cultural aspects of Afro-Hispanic life find themselves at the bottom of a national class hierarchy which emphasizes a common, mixed, ancestry . . . the rise of class consciousness is favourable to the asymmetric ethnic strategies of the new 'blancos' (whites).
>
> (Whitten 1981: 192)

However, as we have noted, the myth of racial democracy has not gone uncontested from its earliest articulation. São Paulo has generated one hundred years of black protest (Reid Andrews 1992), and this culture of resistance has continued in Rio de Janeiro evidenced in the work of the *Movimento*

Negro in the 1945–88 period of Brazilian history (Hanchard 1994). Black resistance is marked today by organizations, cultures and forms of criticism generated within Brazil as well as in Colombia and Ecuador. One of the best known critics of 'racial democracy' is the African-Brazilian intellectual, Abdias do Nascimento, who has provided a revisioning of Brazilian history in which the place of the African-Brazilian population is foregrounded. Do Nascimento is scathing in his attack on 'racial democracy':

> The attempt was to silence millions of Brazilians of African origin with the illusion that, by solving the dichotomy between rich and poor or worker and employer, all racial problems would be automatically resolved. This position of the white Eurocentric ruling elite was taken to the extreme of elaborating an ideology called 'racial democracy' whose goal was to proclaim the virtues of Brazilian race relations, presenting them as an example to be followed by the rest of the world.
>
> (Do Nascimento 1989)

The rhetoric of political discourse is crucial to Nascimento's critique which further reinforces the earlier analysis from Bhabha that 'official' histories exclude the contributions from minority populations which further reinforces their current marginalization. Nascimento correctly resists this story, emphasizing instead the crucial contribution of enslaved labour and demonstrating the intimate bond between racism and capitalist development to the present day. The progressive side of modernity, expressed in the language of democracy, in the development of citizenship rights for which larger and larger numbers of people fought, is found, in fact, to be crucially bound to racism as Nascimento elaborates:

'The enslaved African became a "citizen" as stated under the law, but he also became a "nigger": cornered from all sides' (Do Nascimento 1989: 42). This is part of an ongoing story in which the early-twentieth-century European migrations displaced the African populations, just as in Argentina the indigenous populations were displaced. Now, with the plunder of Amazonia, both populations are again dispossessed.

Theoretically and politically Nascimento's intervention is an important one because it suggests, unlike earlier critiques of 'racial democracy', that there is more at stake than the contradiction between modernity, that is, captialist development and survivals from the early racist plantation economies. Many of the critiques relied heavily on the notion that class would become pre-eminent as the new phase of capitalist development unfolded (Winant 1992). But, given our understanding of the centrality of racism to the modern project, such an analysis, for all its critique of racial democracy, is incomplete. As Winant (1992) notes,

> despite their success at exposing racial inequalities in Brazil and thus destroying the 'racial democracy' myth, the revisionist approaches encountered difficulties when they had to explain transformations in racial dynamics after slavery, and particularly the persistence of racial inequality in a developing capitalist society.
>
> (Winant 1992: 179)

As a counter, Winant suggests an emphasis upon racial formation, 'a process of permanently contested social institutions and permanently conflictual identities' (1992: 192).

Such a view maintains the centrality of 'race' and racism and shifts the focus from economics as the privileged terrain of struggle on to a wider cultural politics. Contestations around national identities, in which the resurgence of racialized and ethnicized politics figures, is one of the key elements of political *abertura*, the struggle for democracy and the re-emergence of civil society. Equally, the state is not absent from this politics but provides both a site for racial projects where a series of discourses and practices racialize populations and a terrain upon which the meaning of 'race' is contested. Nowhere is this more apparent than in the debates around national identity.

In relation to Colombia, for example, 'it is the slippage between including blacks as ordinary citizens and excluding them from the heart of nationhood which characterizes the position of blacks in the Colombian racial order' (Wade 1994: 36).

RACIAL FORMATIONS AND THE 'INDIGENOUS' POPULATIONS

The debates around 'racial democracy' have tended to focus attention on the racism surrounding people of the African diaspora and the colour line between black and white. The racialization of indigenous peoples has been discussed in relation to the notion of *mestizaje*. The idea of 'whitening' is again privileged and is expressed in a language which speaks volumes for the ways in which a Eurocentric view of the world is maintained. As Wade expresses this:

> *Mestizaje* takes on powerful moral connotations: it is not just neutral mixture but hierarchical movement, and the movement that potentially has greatest value is upward movement – *blanqueamiento* or whitening understood in physical and cultural terms.
>
> (Wade 1994: 21)

By contrast, *indios* is a pejorative term used to denote backwardness and distance from the modern project (Stark 1981). It is a discourse which is deeply embedded within the fabric of societies, institutionally and in the commonsense world of daily encounters. The implications of this racialized language for political mobilizations are to be found in the ways in which 'ethnic' identities can be used to promote or suppress political identities. This is clearly expressed by Rigoberta Menchú, Guatemalan human rights activist, who states: 'I am an IndianIST, not just an Indian. I am an Indianist to my fingertips and I defend everything to do with my ancestors' (Menchú 1983: 166).

The language issue has a very specific resonance within the context of Andean cultures resulting from the complexities of the colonial encounter. There is, however, no simple opposition within or between political identities in relation to the ways in which language positions subjects. Indeed, in the struggles between landlords and peasants in Peru, the issue of labelling peasants was integral to the power relations between the two groups. Contestations in the Peruvian *haciendas* foreground the racialization of language and political identities: '*hacendados* (estate owners) could never fix on a single term to define the people living on their properties' (Wilson 1986: 86). They were described economically as workers, paternalistically as children, or as indigenes and so on. Thus, language is not innocent and plays a key role in constructing the shifting terrain of identities which included the pejorative *indios* and its paternalistic cousin, *niños* (children, or childlike). Clearly none of the terms used by the estate owners dignified the ethnic origins of the people involved, and the situation continues today. What Nascimento foregrounds in relation to Brazil and the African Brazilian population is reproduced in modern Peru and the other Andean countries through the adoption of the term *campesino/a* (peasant), a subordinate political-economic class identity with the ethnic referent removed. The language of peasants and workers is central to modernity and the symbolic discourses of nationhood that have been developed in the twentieth century.

Nationalist discourses are generally organized around an account of the nation which privileges a specific construction of ethnicity as part of a racial project in which difference is subsumed. The contours of such national projects are organized by the state which provides legal and administrative definitions of citizens and aliens, insiders and outsiders. As the later discussion in this chapter will show, racialized minorities in Latin America have chosen the terrain of the nation upon which to struggle, entering into a debate which is currently having a profound impact on the ways in which nations, national identities and nationhood are thought and practised. Before we explore this further we need to further unravel the meanings and expressions of racisms in relation to our conceptual framework, elaborated in Chapter 1.

RACISMS: WRITTEN ON THE BODY

I come from a family in which some have been *cholos* and others *indios*. . . . We have something ethnic, where we originate . . . and you have to take the physical side into account. I belong to the indigenous ethnic group, but I feel more Spanish. So in effect (*de hecho*) I'm a *mestizo*!
(Authors' interview with a man of 80 in an upper-working-class neighbourhood of Quito)

In conceptualizing racism we are indebted to Winant (1992; 1994) and his theorization of racial formation and racial projects and to the work of Hall (1990, 1991), Bhabha (1994), Gilroy (1993) and Rattansi (1994) in their explorations of the articulations between racism, cultures, identities and nation. Equally the work of Michel Foucault, despite the criticisms generated by Homi Bhabha, does offer an important and suggestive way of understanding racism as a regime of power that comes to rest on the body, and in which subjects are made both visible and invisible simultaneously as 'disciplined' subjects. Racism is, therefore, not unitary or fixed but organized around different signifiers, most especially those written on the body, that privilege biology in one moment and culture in another, the consequences of which are violence (symbolic and physical), Otherization, inferiorization, exclusion and subordination. Thus, Gilroy (1993: 22) argues: 'there is no racism in general and consequently there can be no general theory of race relations or race and politics'. And he continues: 'Races are not, then, simple expressions of either biological or cultural sameness. They are imagined – socially and politically constructed' (Gilroy 1993: 20). This latter point has major implications for the constructions of 'fictive ethnicities' in relation to the imagined communities of Latin American nations. However, ethnic identities 'real' or imagined, like class identities, are always constructed and refracted through gender identities (explored in Chapter 6). Writers like Rattansi (1994) and Gilman (1991) have also emphasized the role of sexualities in the processes of racialization and this sexualized racism is explored in the study by Guy (1991), for example, of late-nineteenth-century prostitution in Buenos Aires, where Jewish male migrants were constructed as the perpetrators of a 'white slave trade' in women from Europe at a time when the Jewish populations were persecuted across Europe and the Baltics. These discourses, which privileged the racialized body as a signifier of Jewishness, were an attempt to fix the Jews, most especially the male Jew, morally and physically within a disciplinary mode which constructed Jewish people as marginal to the imagined community of the nation, shared moral values, throughout the nineteenth and twentieth centuries, although the roots of this lie in earlier epochs.

The theorization suggested moves us away from simple binary accounts of racist/non-racist, black and white and so on (Rattansi 1994). Instead, and most especially in the context of Latin America, we need to find ways of engaging with the malleability of racisms. Thus, the debate around 'racial democracy' has tended, not surprisingly, to focus on the relations between black and white people and to be organized around discourses of colour and the specificities of the negro body. The historical legacy of enslavement and the sexual fantasies of Europe collude in the generation and sustenance of forms of racism which are expressed and experienced as tied very specifically to the body. As Winant (1994) expresses this

> so we might usefully think of a racial *longue durée* in which the slow inscription of phenotypical signification took place upon the human body, in and through conquest and enslavement, to be sure, but also as an enormous act of expression, of narration.
>
> (Winant 1994: 21)

One exploration of this narration is contained in the analysis of Xuxa, the phenomenally successful Brazilian presenter of a children's programme now exported to the USA (Simpson 1993). Xuxa both embodies and represents the blonde, white woman at the apex of the colour hierarchy with all the symbolic capital that implies. During her career she has used her whiteness to re-invent herself as a cultural icon in Brazil, matched only by African-Brazilian soccer legend, Pelé (her one time lover). As Simpson notes,

Xuxa's claim to Brazilian identity is made plausible by that measure of belonging at the same time that she displays the features of the coveted other, the blonde that is so often viewed as a symbol of superiority. Thus image management gives Xuxa the flexibility to play on Brazil's mental habit of exalting the blonde while simultaneously asserting kinship with a nation of nonblondes.

(Simpson 1993: 15)

Similar accounts can be found in Parker's (1991) explorations of carnival and popular cultures in relation to the black body and sexualities and, in the examination of Brazilian Capoeira, the martial arts/dance and song form with its roots in enslavement and the African diaspora (Lowell Lewis 1992).

The embodiment of 'race' and its reconstruction within racism is clearly expressed by African Ecuadoreans who in discussing racism were quite clear that it is their skin colour that people see first and which is the crucial signifier for their elaboration and Otherization in the eyes of Ecuadoreans. In the responses to interviews in the Ecuadorean coastal cities of Esmeraldas and Guayaquil, it was racism and poverty as the defining features of black people's lives in Ecuador and Latin America that were emphasized, but not exclusively. There was also a discourse on the power of black people as sportsmen and women and the contribution that African cultures have made to the past and present of Ecuador. Racist incidents are part of everyday life for black people in Ecuador and one African Ecuadorean woman, Sonia, recalled an encounter in downtown Guayaquil. A white man had stepped on her foot and she looked to him for an apology, but he replied with, 'Get off, get back to the kitchen and the whorehouse where you belong.' This painful encounter foregrounds the ways in which commonsense racisms are constructed within the articulation of gender, sexuality, 'race' and class with the body as signifier of racialized, gendered labour power and sexuality.

Similar articulations re-emerge in the intermezzo world of *mestizaje*, which has provided a mythical version of indigenous–white relations in which both are equally valued and hybridity is seen as the outcome of democratic intermingling. The historical basis for this is dubious and includes the genocidal practices of the West. Indigenous peoples have been massacred in a scramble for land and domination in Latin America, either early in the colonial period for the heartland of the indigenous empires or, more recently, in the Amazonian area. Although appearing as merely historical, these processes provide a series of themes today recycled in order to generate a racist discourse which placing indigenous peoples as subjects/objects in specific ways, bound to earlier forms of 'Otherization'. In the current period it is the signifier of culture, often written on the body, which organizes racism against indigenous peoples contributing to the ways in which they are subordinated, excluded and constituted as 'the ultimate other'. Forms of racialization and racism do not go uncontested; the indigenous people's movement has shown both strength in numbers and strategic acumen by using the confluence via culture between 'race' and nation as a terrain which they have retrieved as a site of struggle and strength – an issue to be further explored later in this chapter.

In summary, racisms are plural, drawing on the past but always of the present. Racisms are specific and contingent, thereby allowing for the ways in which those who are subjected to racism can themselves usurp the symbols around which racism is organized. Thus, the black man uses his physicality to demonstrate prowess on the soccer pitch and the indigenous trader from Otavalo uses the signifiers of culture (long hair for men and dress) to signal ethnic authenticity as part of a strategy for successful capitalist development. It is not surprising, therefore, that African Ecuadoreans interviewed in Esmeraldas emphasized the racism of Latin America alongside the power of black people and their historical contribution to the nation-state of Ecuador. Equally, African Ecuadoreans recognized and celebrated the power of indigenous peoples, especially in relation to their current political organization. Discussions with African Ecuadoreans in Guayaquil and Esmeraldas were highly politicized, emphasizing power relations as expressed in economic and political terms but always as racialized relations. A number of people we interviewed are members of the African-Ecuadorean Association and of political parties with Leftist agendas. While acutely aware of the complex relations between racism, ethnicity, gender and class, the modality within which these articulations are understood is clearly that of racism. What is important is that this is a popular understanding of racisms as regimes of power organized both through discourses and discursive practices that impact upon, and organize, black people's lives in any given space–time configuration.

The importance of grounding regimes of power in specific sites in relation to racism is theorized within the work of Winant (1992, 1994) through his use of the notions of 'racial project' and 'racial formation'. These conceptual tools invoke the state and its relations with civil society in ways which allow for the de-centring of the social, which is consistent with our understandings of nations and national identities. Winant (1994) moves this theorization further in his conception of a 'decentred hegemony' which allows for a racial hegemony to be fractured and reconstituted via the diffusion of powers and sites which are both within nation-states and trans-national. Equally, it could be argued that the crucial element of 'hegemony' lies in the understanding that hegemony is de-centred. Consequently, racisms can be organized into 'racial hegemonies' and racial projects which promote the national and the centred nation. However, racisms simultaneously disorganize the national through the fractures and divisions that are promoted and sustained via racisms, generating instead the de-centred nation. Winant wishes to emphasize not so much the complex structural processes that generate consent but the understanding that 'an effective hegemony constructs its own subjects' (Winant 1994: 269) as well as the counter to these 'official' subjects in the subversions of identities. It is precisely in this realm of identities and their politics that Winant wishes to invoke a post-structuralist understanding of the subject and significations as de-centred. One of the crucial ways in which these configurations are expressed and theorized is via discourses, the narratives around state, nation democracy and citizenship and it is to an exploration of these interlocking fields and subjectivities that we now turn.

NATION, STATE AND CITIZEN

To foreground 'the nation' as a terrain of struggle and contestation as the indigenous people's movements have done is one part of the ongoing politics of national identities. Such a politics involves taking on the state that organizes the account of 'national' and 'alien', but it also invokes struggles around democracy and citizenship. These are not new issues but their articulation within an anti-racist framework that is deeply embedded in the social movements generates new discourses and forms of organizing. Consequently, the founding statement of the Latin American Subaltern Studies Group (1993: 112) states: 'the force behind the problem of the subaltern in Latin America could be said to arise directly out of the need to reconceptualize the relation of nation, state and "people".' The context for this political reframing, however, is itself a contradictory one in which a re-visioned democracy is 'allied' with free market economics and the power of the dollar. What this has made possible is an arena into which social movements can step and offer an alternative account of democracy and citizenship but within an economic programme known as 'liberalization', renowned for its illiberality in relation to civil rights. There is in process the generation of new hegemonies within the different states bound to the rearticulation of 'the nation'. Mallon's work, for example, suggests

> a Mexican state that emerged as hegemonic because it incorporated a part of the popular agenda and a Peruvian state that never stabilized precisely because it repeatedly repressed and marginalized popular political cultures.
>
> (Mallon 1995: 311)

It is interesting to reflect on the ways in which the African Ecuadorean people we interviewed responded to the question on the meaning of citizenship in Ecuador. Almost all the black people in Esmeraldas saw citizenship as obeying the law and carrying the right documents. As one African-Ecuadorean man commented, citizenship meant: '*tener todas sus documentos en regla*' (having all the right documents). At the same time, however, when asked 'Who has power in Ecuador?', the answers spoke of a different, less disciplinary, conception of state–citizen relations because the responses coalesced around 'the indigenous'. The answer is a political one, in the same way that it was for those who replied, 'black people'. It demonstrates clearly the importance of racial politics to the development and consolidation of democratic forms in Ecuador, as elsewhere in Latin America (see Winant 1994; Mallon 1995). The reference to the indigenous peoples is not, however, an ethnically descriptive category but is generated by the power of the *Confederación de Nacionalidades Indígenas del Ecuador* (CONAIE: Ecuadorean Confederation of Indigenous Peoples) and the ability of the Confederation to mobilize large numbers of people around issues of national identity, culture and citizenship. The political base for this work is highly differentiated but draws on the historical and contemporary experience of 'Otherization' and from this a space in which to generate a collective subject as part of a strategy for change.

This politics is being framed within the problematic and uneven 're-democratization' of the Latin American states. Of the twenty republics of Latin America, one can see in their past 'the vanguard of international liberalism when they repudiated monarchism, aristocracy and slavery in the past century' (Whitehead 1992: 312). Following this repudiation were a variety of oligarchic, military and authoritarian populist regimes. As Rouquié (1987) notes, in 1980 two-thirds of the peoples of Latin America lived under a variety of forms of military rule, from the limited militarization of Colombia to the inclusionary modernization of the Peruvian military in the late 1960s with its agrarian reform. In Ecuador, General Rodríguez Lara came to power in a coup in 1972 and proclaimed the new government: 'revolutionary, nationalist, socialist-humanist and in favour of autonomous development' (quoted in Rouquié 1987: 328). The state controlled the most important resource, petroleum, and modernized agriculture with the introduction of wage labour and a new road network but imports and the bureaucracy grew while business interests became restive. In 1976 General Lara was ousted by the chiefs of staff committed to a return to civilian rule. Strong on nationalist rhetoric, this period in Ecuador's history demonstrates the relationship between economics, the nation and the military and the ways in which, in Ecuador, for example, these relationships are constantly reinvented.

Whitehead suggests that neither populism nor repression have delivered quiescent societies, nor the conditions for economic progress, generating instead a kind of truce from which the new democracies have emerged. He calls this 'democracy by default' (1992: 314) and 'façade democracy' where civil power and the rule of law are not in control of the military, although there are more and more moves in this direction. The consolidation of democratic forms is slow to emerge and instead there are powerful forms of 'presidentialist' government. At the same time free-market economics, encouraged by international institutions such as the World Bank, are having a powerful impact on the democratic paths taken by individual states. Chile is often cited as the country that has progressed furthest along this path with growing economic strength and a measure of consolidation, yet the path is fraught with problems most especially the ways in which neo-liberal economies demand and make use of a lack of freedom within civil society. As Whitehead notes:

> Certainly participatory democracy tends to clash with the ideal of a 'depoliticized' market system, and far from advancing such causes as women's rights, racial equality or a social rights conception of citizenship, the neo-liberal model tends to dismantle whatever protections may have previously existed.
>
> (Whitehead 1992: 320)

Despite these shortcomings, the alternative of 'unstable populism' may be worse especially for the poor, women and civil rights. In many discussions on this theme, it is liberal democracy that is held out as the model, a model with its roots in the modernity of the West, and that 'democracy by default' is the present outcome (Fukuyama 1993; Watts 1993). However, there is no linear progression towards a model that is constantly reorganized by its conditions of existence. Instead, Latin America, with its great diversity and resilience expressed in popular cultures and through civil society, also has the possibility of re-visioning notions of democracy that draw in questions of ethnicity and gender in ways that are only beginning to be part of the discussion elsewhere. This is a point of view expressed forcefully by Yúdice (1992) who notes:

> Universalizing modes of democratization have not been the most successful in Latin America ... due in great part to the tendency to understand democratization in terms of modernization. ... The problem is, of course, that modernization has severely handicapped many groups who hold to these traditions.
>
> (Yúdice 1992: 23)

Central to this ongoing discussion has been the work of Laclau and Mouffe who provide both a critique of Marxism and liberal democracy. Against the class-determinist socialist position, Laclau and Mouffe (1985) argue that the muliplicity of interests and subject positions cannot be reduced to class struggle. Thus, against the binary of the basic economic contradiction of capitalism as it was defined, Laclau and Mouffe insert the plurality of antagonisms and a fractured politics which is part of the disaggregation of the social, the de-centred social in what we have come to understand, whether we like the term or not, as post-modern societies. Thus, there can be no simple unitary goal for politics nor a linear progression (as the Enlightenment project) suggested towards progress and within democratic liberal discourses towards liberty. There are many liberties and a multiplicity of political goals and identities and the ways in which these multiplicities are forged into collective subjects is itself the work of politics as, of course, Antonio Gramsci understood so well. The psychoanalytic notion of interpellations is important in this reframed politics offering subject positions which are not only embedded in class or that belong simply to a class. In this sense so-called 'bourgeois democracy' can, therefore, be freed from the liberal democratic traditions of representative democracy and be re-cast within a radical democracy. Such a vision of post-modern politics invokes the 'new' social movements as the key collectivities of political action within an account of power that, following Foucault, is seen to be diffuse and contingent. Such a vision brings immediately to mind the states of Latin America where, despite the ferocious attacks on civil society and political opposition, the collectivities of struggle remained alive (initially in clandestine ways) allowing a form of civil society to be in place when 'democracy' was reinvented.

Chantal Mouffe's earlier volume (1992) addresses more specifically the issue of citizenship and democratic forms but as she notes, 'Our understanding of radical democracy, ... postulates the very impossibility of a final realization of democracy' (Mouffe 1992: 13). The lack of closure as a defining feature of democracy underlines the importance of the principles of liberty and equality and that these are, ultimately, unrealizable. This ensures the indeterminacy of democracy and, thereby, its radical potential. Within this the social movements are prime movers in shifting the terrain of struggle within democracies or towards greater democratic freedoms, away from the corrupted and institutionalized parties that have domesticated the Left radical vision of politics in Latin America. As we write the PRI (Partido Revolucionario Institutional or Institutional Revolutionary Party) – the name itself speaks volumes – in Mexico have again been returned to power, a hegemony exercised over the democratic process in Mexico since 1945. Dr Zedillo, the leader of the PRI, described the election day as 'exemplary' and said there are 'no losers – the most important victory is that of the Mexican people' (quoted in the *Guardian* 23 August 1994). Against this corporatist notion of party and people has been the recent eruption of a re-visioned Zapatista revolt by poor peasants in one of the most deprived areas of Mexico. Thus, what appears at first acquaintance to be a settled regime, one 'that emerged as hegemonic because it incorporated a part of the popular agenda' (Mallon 1995: 311), is not, especially in relation to the issues of ethnicity, poverty and land – a contentious articulation throughout Latin America.

As the previous section has suggested, the historical significance of issues around ethnicity have again become key moments in the politics of the Latin American states since the return to 'democracy'. Democratic political rhetoric and practices have provided a series of sites in which contestations can be expressed and organized. Thus, the social movements have ceased to be small-scale or even national phenomena and have become, especially in relation to the indigenous peoples, a trans-national collectivity. The success of the indigenous people's movement in Ecuador, linked to movements elsewhere, in Brazil, Peru and Colombia, for example, is part of this story, and one which demonstrates the power of social movements within a specific conjuncture. Despite the processes of 'modernization' under a neo-liberal regime which looks ostensibly like unpromising terrain for emancipatory projects, the movements have grown since the mid-1980s. There is, however, a complex and contradictory story of competing accounts of modernization. The military in Ecuador do not support privatization financed by foreign capital on the grounds that it is against the national interest; however, the bourgeoisie and its political parties also have a rhetoric of 'the national interest' bound to economic development through privatization and foreign capital which secures the personal interests of this particular section of the population. The contradictions came together in Quito in 1994 with a public demonstration in opposition to the privatization of power supplies where workers, the military and a multiplicity of groups were in an alliance against the government.

The complexities of the Ecuadorean situation highlight the articulations between state, nation and citizen. Equally, the current politics of the social movements has shown strategic acumen but have also claimed a representational space at the symbolic level. The politics of the social movements brings together the 'dispersal of the social', to use the phrase from Laclau and Mouffe (1985), with the complex ways in which collective subjects and the politics of identities are generated and sustained. Within the ever-growing literature on the social movements in Latin America, Alvarez and Escobar (1992) raise precisely the issue of collective subjects and the politics of identities and the question (similar to that asked of the new democracies in Latin America), what is the life and impact of the social movements?

The work of Melucci (1989, 1992) is helpful in unpacking this question. In his more recent work he reiterates and develops many of the themes of his earlier work, which sought to examine the ways in which in an increasingly fractured world meanings are developed that provide a basis for political action. In the Latin American context this is part of 'those rearticulatory practices that seek to assume alternative traditions within modernity. These involve the struggles for interpretative power on the part of peasants, women, and ethnic, racial and religious groups' (Yúdice 1992: 23). Tied to these conceptions of 'rearticulatory practices' is Melucci's attempt to undermine the ways in which the efficacy of social movements is judged in terms of material gain rather than symbolic outcomes or symbolic capital. For it is at the level of the symbolic that the social movements are most successful and most powerful. In his more recent work, like Nederveen Pieterse (1992), Melucci also takes issue with the understanding that social movement politics is 'liberatory' with a single undefined goal of 'liberation' or 'emancipation'. Against this view he argues for the complexities of current politics and the importance of the cultural field. For it is the ways in which the social movements enter and engage with the cultural field and seek ways to subvert and shift the cultural codes which is crucial to their politics. Thus, the symbolic and the imaginary are privileged in the politics of social movements. As Melucci notes, 'when power is concerned with the control of cultural codes, the main role of social movements is that of making power visible and of opening up civil society as the public space for societal debates' (1992: 323). What this includes, of course, is the deconstruction of the binary public/private so that issues of sexuality, the politics of the body, like reproductive rights, can become part of the arena of public debate. Although in Ecuador, like other Latin American countries, these issues have remained predominantly underground and outside the public domain (see Chapter 6 for a more extended discussion).

Rather, it is the issues of ethnicity and culture within the terrain of national identities that has proved so productive of new ways of seeing society and the state. Perhaps this is in part due to the failure of official discourses on the nation to become part of the commonsense. In Ecuador, for example, Rouquié (1987: 25) notes: 'when Ecuadorean Indians were asked what the notion of *patria*– fatherland –meant to them, they answered – after a century and a half of independence, national symbols, and patriotic affirmations – "A bus company".' (There is a company with that name in Quito.) As a counter to this lack of 'collective imagination' the success of the indigenous movement in Ecuador, and particularly CONAIE, can be seen to relate to a willingness to acknowledge that a collective subject must be crafted out of the diversity of indigenous peoples. Languages, localities, customs, traditions and fiestas are all very specific and yet it has been possible to weld together an effective opposition to the status quo and to envision a new role for indigenous peoples, a new place in the nation. In part this relates to the importance of land which is a unifying force and thus, the materiality of the struggle is deeply embedded in the economic and cultural forms of the people involved, and organizing around land rights has been imperative, an issue explored in greater depth in

Chapter 5. At the cultural level there has been the struggle over language and bilingual education, yet this is more fractured due to the variety of languages and the tendency to use Quichua as a shorthand for all indigenous languages. There has been some success in the cultural field, directly linked to the issue of shifting cultural codes and claiming cultural capital from what has been previously marginalized. Latterly too the language of racism has entered the discourses of CONAIE with the understanding that indigenous peoples are one part of this story and that people of African descent also share a history of exploitation and racist abuse.

What is of special interest in relation to the indigenous movements is the seeming contradiction between 'claims to authenticity' in relation to the history and culture of Latin America and a political agenda which has all the demarcations of the post-modern (see, for example, the discussion of the Chiapas rebellion by Burbach 1994). If as Ross (1988: vii) states the marks of the post-modern are that it is 'decentred, trans-national and pluralistic', then CONAIE bears all the marks of a post-modern politics in which the organization is a confederation but with a widely diverse group of peoples within, and it is trans-national in orientation while it is dealing with issues within the nation-state. Pluralism is the key to its platform in relation to issues of citizenship and national identity framed in the call for Ecuador to be a 'pluri-nation' and for the constitution to change in order for this to be recognized. In discussions with CONAIE it was very clear that here was a politics organized around the notions of fluidity and contingency that is proactive in wresting space within the political field. CONAIE has used strategic uncertainty to its advantage, refusing to be drawn into conventional alliances but defining allies and enemies from its own standpoint. It is in this sense that CONAIE has been able to galvanize a wider opposition especially since the elections of 1994, that returned a Christian Democrat majority and routed the Left. CONAIE was prepared for and able to step into the vacuum left by the death of the conventional opposition, while simultaneously showing a willingness to engage in dialogue with the military at the highest level. What is currently appearing in Ecuador is the possibility of a new settlement in which the centre and the margin have allied in relation to specific issues at some distance from the neo-liberal agenda. Both the military and CONAIE are able to mobilize around a political rhetoric which is strong on populism and the future well-being of the nation of Ecuador.

Claims to authenticity are shared by both the indigenous and Afro-centric movements of Latin America. In Ecuador, as elsewhere, there was a clear bifurcation between the claims to an African identity among black people with a 'roots story' located in the interface between an imaginary Africa and the historical and cultural legacy in everyday lives – through folkloric elements, religious and therapeutic practices, music and dance – and urban black culture favoured by young black people who watched the film *Malcolm X* with relish, listened to Ice T and wore street styles from New York. In contrast, an indigenous identity drew the past and the present together as a symbolic act of enunciation through language, clothes and folkloric elements which are part of everyday life. It was, for those who were politically active, a self-conscious reinvention of 'indigenousness' made public through the throwing off of Christian names in favour of indigenous names, practising bilingualism through self-taught Quichua and programmes of education in alternative ways of knowing and being in the world. These were vitally important symbolic acts foregrounded in the politics of representation for CONAIE members and crucial to the generation of the collective political subject. But, it was not a retreat into authenticity; consistent with a post-modern politics this was a refashioned identity constructed in relation to skills essential to the struggles within a modern, disciplinary state and which required the legal acumen of Luis Macas, the CONAIE leader, and an ability to use the media.

The Latin American social movements are generated from within civil society and operate on the terrain of democracy; by so doing they throw into sharp relief the limits and possibilities of a citizenship defined in relation to liberal or representative democracy, the crux of which is the notion that via the ballot box interests are 'represented'. At the base of this model is a notion of power as finite and divided between those who have power, the power-ful, and those who do not, the power-less. Clearly, Foucault's work has recast our understanding of power as fluid and relational and coalescing in specific sites. As the politics of the social movements show, power and resistance constantly redefine where power lies and what it produces. Central to the liberal democratic view is the notion of the abstracted and atomized 'citizen' in whom difference whether of gender, 'race', sexuality or class are not recognized. The social movements, organized as they are around the very axes that this version of the citizenship makes invisible, challenge this account from within its own terms and offer a more contextual, multiple and contingent construction.

In Latin America the role of women in social movements, and at the forefront of the deconstruction of the separation between public and private, is well known, and in this book further explored in Chapter 6. What has tended to be absent in many of the discussions of gender and citizenship, for example, Walby's (1992) recent critique of Marshall, are the issues of racism and ethnicity. This is in part due to the ways in which citizenship tends to be separated from issues of nationality and the nation. It would appear to be an extraordinary 'oversight' but it is, in fact, part of the way in which the modern and its political institutions are seen as ethnically neutral and formalistic, beautifully expressed by Gabriel Garcia Marquez in relation to Colombia:

> I believe that we are acting, thinking, conceiving and trying to go on making not a real country, but one of paper. The constitution, the laws . . . everything in Colombia is magnificent, everything on paper, it has no connection with reality. . . . There is a democratic tradition, repressed a long, long time ago, which is the only hope for us, for Colombia.
>
> (*Semana* 20 April 1989)

It is precisely this repressed democratic tradition that is re-engaged in the debates around radical democracy, that foreground pluralism and the complex configurations of identity politics in democracy. As Mouffe (1992: 4; 1993; 1995) notes, 'A radical, democratic citizen must be an active citizen, somebody who *acts* as a citizen, who conceives of herself as a participant in a collective undertaking'. Thus, what is crucially at issue is the way in which citizens 'are constituted only through acts of identification'(Mouffe 1992: 11). It is this process which offers the possibilities of a radical democracy that extends beyond class boundaries and identifications to include the social movements. This pluralist conception is consistent with the notion of multiple selves and political action in the context of the de-centring of the social, 'identity is not what one is but what one enacts' (McClure 1992: 124). What is at issue here is the way in which contingency produces a community of interests enacted via a political identity which is not immutable or essential and is, therefore, beset by tensions and contradictions. These tensions and contradictions are themselves transformed in the process of action, the definition of interests and allegiances. This is a much more dynamic account of the contestations within politics and offers ways to theorize the encounter between diverse interests that are constantly overlapping and realigned. Included within this de-centred politics is the arena of the state which we have previously argued is not conceptualized as unitary with 'interests, but as a series of overlapping sites within which politics becomes a strategic and contingent terrain' (Westwood and Radcliffe 1993). Against a view of fixed positions that are occupied or vacated in the state, an understanding of sites, diffusion and relational powers are part of the practical politics of the politically active in Latin America. For example, CONAIE and its dialogue with the upper echelons of the military has proved an important weapon, yet does not obscure the power of the military and its control over the means of violence, a powerful component of national cultures throughout Latin America. The more communitarian thrust of the republican view has important implications for a radical democratic construction of citizenship, which is located with the equivalence of the identity of citizen but not one that is foreclosed (Mouffe 1992).

The making and remaking of democratic politics in Latin America is the terrain in which the social movements have been key players both in terms of defining the rules of political engagement and in terms of constructing political subjects. Central to the political projects of the social movements is the active citizen and the notion of the citizen as a political identity in process of generation and sustenance. This has important implications for our concerns in this book because the citizen and the national are constantly intertwined and this is clearly expressed in the current literature from Latin America, some of which has been discussed in Chapter 1. Equally, the issues of racism with which we began this chapter are foregrounded in relation to the articulations between racial formation and democratic cultures, expressed in the public debates and mobilizations by African Brazilians against the myth of racial democracy. Racisms and the processes of racialization contribute towards the de-centring of the nation through difference, the exclusivities that are the hallmark of racial categories, and the politics which is organized against this. At the same time, narratives of national identity reinvent not only histories and cultures but also mythical lineage through blood that is written on the body and expressed in terms of a 'fictive ethnicity' which is vital to the centring of the nation and expressed in a myriad cultural and legal forms.

Albó (1993), in opening a discussion on post-modernism in Latin America, foregrounds 'the national question', understood by politicians as a potent mobilizing force but under-theorized by social scientists who have yet to fully explore the ways in which the state organizes and monopolizes 'the concept of the nation'. It is not surprising that the growing power of indigenous groups should decide to counter this state monopoly by, as Albó suggests, 'a growing convergence toward the identification of their projects as national' (1993: 24). However, the conception of the national is a very different one from the exclusive loyalty demanded by national identities as they have been understood historically. Instead, the conception of pluri-nationality invokes the equivalence of multiple loyalties within a de-centred social. In this sense it works against the centre/margin binary and a simple unitary view of the nation and the power of the state within this, contributing to a vision of Latin America which offers a recognition of difference and the development of a 'multi-national' identity which provides spaces for ethnic and regional diversities. Taking it further, suggests Albó, invokes a United States of Latin America which will not foreclose future developments but sustain independence and the vitality of the region's diversities. Such a political vision is, of course, mindful of the global interests of multi-national capital and its relations with those elements of indigenous/national capitals.

These new formations and visions, suggests Albó (1993), are easier to see from below or above. That is clearly evident in the development of CONAIE, which has a national agenda within Ecuador but is committed to a transnational agenda in relation to commonalities across borders and states and modes of organizing. At the same time local/regional and ethnic/cultural spaces and loyalties are of prime importance and demonstrate the diversity within the collective subject of CONAIE. Complex identities at the local and regional level, within ethnic and language groups, are a major component of the national throughout Latin America which, seen in this way, generates and sustains a myriad 'little nations'. In part, these are sustained by the forms of popular nationalism that are part of everyday life and are further explored in relation to Mexico and Peru in the detailed analysis of regional communities and forms of communal hegemony and alternative nationalisms by Mallon (1995). The importance of these pluralist conceptions and re-visioned politics is expressed by Mouffe (1995: 265), 'a democratic politics informed by an anti-essentialist approach can defuse the potential for violence that exists in every construction of collective identities and create the conditions for a truly "agonistic" pluralism'. Latin American states which have known the violence of exclusivities and Otherizations are precisely the terrain on which an 'agonistic pluralism' can be generated, but, not to labour the point we would suggest that the arena of the national and national identities, curiously neglected by both Laclau and Mouffe, will be crucial to this re-visioned democratic politics. In Chapter 3 we explore more fully the 'making of the nation' through an analysis of official discourses and practices embedded in the social formation of Ecuador.

3

ECUADOR

Making the nation

The Ecuadorean state generates a conception of Ecuadorean identity, one grounded in what is supposedly the uniqueness and value of *ecuadorianidad* (Ecuadorean-ness). When looking at official Ecuadorean nationalism, it is useful to consider three key 'fields of power' (the 'sites') of history, territory and population around which discursive constructions of nationhood take place. In Anderson's discussion of Asian official nationalisms, he suggests that

> [nationalisms] imagined [the nation's] dominion – the nature of the human beings it ruled, the geography of its domain, and the legitimacy of its ancestry.
>
> (Anderson 1991: 164; see also A. Smith 1991)

Accordingly, the chapter is divided into three sections around which practices and discursive material from state sectors are organized, as the official practices and discourses around the issues of history, territory and population are broader than specific practices of census, map and museum. Ecuadorean official national discourses and practices are analysed in terms of these sites, to illustrate general Latin American official discourses. These three dimensions of official national constructions and citizenship cut across the thematic organization of the state into sectors dealing with say, education, the military and political decision-making.

HISTORY

As in other Latin American countries, the creative narration – and in some cases invention – of histories to suit nation-building purposes is found in Ecuador (cf. Hobsbawm and Ranger 1983), and 'nationalism too invents dates' (A. Smith 1986: 177; Szászdi 1963). Rather than being restricted to academic history and officials' discussions of events, national histories spill over into a variety of spheres, including school education, commemorative dates and history embodied in architecture. By examining the process of performing and displaying national history in Ecuador, this section notes a contradiction in the aims of national history, as it projects itself back into a mythical, heroic past and forward into a golden future (A. Smith 1986, 1991; Kandiyoti 1991a).

Teaching and learning histories

Regardless of the organization running a school (which can be religious, military or the secular-state), the curriculum is the same across Ecuador, diffusing official national discourses. In this respect, the guiding principles of education are those with which every school child comes into contact. According to a spokesperson, the goal of education is to develop the cultural inheritance of Ecuadoreans and to maintain and preserve all the values of nationality.[1] With teachers most likely to be trained in one of a limited number of national colleges, the values expressed in and through education are remarkably uniform.

The writing and teaching of official versions of history are key elements in a nationalist education process, relying on more than the timetable slots for 'history'. In Ecuador, the organization of the school week is around a 'civic moment' which recalls key historic moments to remind students of them. The events recalled during this civic moment are largely connected with the historical partitioning of territory claimed to be Ecuadorean. Utilizing the idea of the nation as a (social) body, one education spokesperson remarked,

> It's necessary to know the history of Ecuador. Ecuador has had many problems of dismembering [*desmembración*] of territory; as a consequence this territorial dismembering has . . . led to the near disappearance of the state for historical reasons and problems which are not only territorial, but to do with the hegemony of some Latin American nations.[2]

Civic education, a compulsory part of schooling for students to the end of secondary school, brings together the heroic struggles of the national past, making them components of *contemporary* national self-imaginings. Not stable and unchanging, these identities are liable to shifts over time as the role of nation-building draws the state into reinventing its history to fit with contemporary demands. For example, in Peru, the Inca empire gained an increasingly favourable image in textbooks (Portocarrero and Oliart 1989).

In Ecuador, there are tracings of 'national' history to the pre-conquest period, although rather than refer to the Inca past (which would tie the country 'too closely' to Peru, its rival in the south) it refers to the (mythical) kingdom of Quitu. Drawing upon the writings of Padre Juan de Velasco, an early chronicler – who invented several dimensions of 'Ecuadorean' history (Brading 1991) – school history places considerable emphasis on the Kingdom of Quitu and the ethnic groups which comprised it. Early secondary textual narratives trace the history of culture groups linked through marriage or conquest in (what later becomes) the national space. Resistance of the Quitu 'sovereigns' to Inca conqest is also prominent; by projecting on to the screen of the past a recent event (the Peruvian invasion of Ecuadorean territory), the history curriculum reminds children of the southern threat. Curricular history later returns to issues of territorial sovereignty, looking at the several colonial edicts and post-colonial treaties shaping Ecuador. The conceptual framework used in these histories is in many respects a Marxist one, explaining modes of production, colonial exploitation of indigenous labour, and colonial class structure (e.g. García Gonzalez 1992). History curriculum policy for Years 6, 7 and 8 outlines how students should learn about the distribution of historical ethnic groups, as well as about indigenous uprisings against colonial and republican rule. School history lessons thereby integrate notions of human rights and the new indigenism (see p. 69), into their pedagogic material.

Every Monday morning a ceremony called the 'civic moment' takes place in schools bringing together the national flag, the national anthem and historical events. According to an education minister,

> For several years now, many years, we have been developing or encouraging the civic spirit by means of what is called the civic moment.[3]

Pupils (from the first grade of primary school) and teachers come together in each school to raise the flag, sing the national anthem, and then

> they briefly commemorate some civic or national happening [*aconte-cimiento*] which relates to the calender of that week. This is done because . . . [pause] it is necessary to know the history of Ecuador.[4]

The round of national time begun at school then continues in national life at public occasions, similarly commemorating key dates in the country's history.

Secular and national time: the annual progress of nationalisms

Nationalisms commonly provide secular dates whose appearance marks time in the secular-nationalist annual round, reinforcing the conceptualization of continuity and stability in the national space (Anderson 1991; Gillis 1994). With the separation of Church and state (in many Latin American countries as in Ecuador), secular dates provide the major 'national' markers of time, although in addition Catholic religious dates feature. By reminding citizens of key moments in their history, the dates complement and continue the process of recording and remembering significant dates which starts at school with the 'civic moment'.

The most important date in the state calender – and certainly the one with the most 'national' connotations – is the 10 de Agosto (10 August) – Ecuador's 'national' day. Marking a (failed) attempt to break from the Spanish crown in 1809, 10 de Agosto is now widely celebrated and remembered as *el primer grito de la independencia* 'the first shout for independence'. In other words, failure in defining national identity is converted into the primary post-colonial date, resonating not only with 'internal' national significance but also with reference to other Latin American countries: Ecuador is presented as *the* precursor of Latin American anti-colonial struggles. On 10 August after general elections, presidents and their governments are inaugurated, while in other years presidents may give major public addresses. President Galo Plaza for example chose to speak about the unacceptable Rio Protocol on the 10 August 1951 (Molina 1994: 162). On a more *ad-hoc* basis, other days can become the focus for official nationalist discourses when there is a percieved need to 'remind' citizens of a 'forgotten' past event (Renan 1990). Such celebration may also be associated with the unveiling of a monument or an official visit to an historic site, thereby grounding historical events in actual and (via the media) visible contemporary places.

Architecture and nation

As AlSayyad suggests, buildings help 'serve to establish national conscious-ness' in both colonial and post-colonial times, while national ideologies can draw upon buildings from a wide range of eras to create a unique national place (AlSayyad 1992). Public monuments to key figures and events in national imaginations can 'map out' a national territory, as effectively as national maps. By providing key reference points and material markers on the ground of the nation, monuments provide nodes for feelings of belonging, identity and continuity (as well as of alienation, difference and distance, as seen in later chapters) (A. Smith 1986, 1991; Anderson 1991; Daniels 1993). Certainly, the post-colonial state of Ecuador draws upon the rich nationalist symbolism of buildings and cities. The Ecuadorean Instituto Nacional de Patrimonio Cultural (INPC: National Institute of Cultural Patrimony) is responsible for preserving and protecting the colonial townscapes of key 'historical centres'. By the early 1990s, the cities of Quito and Cuenca were declared as historic centres, requiring municipal ordinances to protect the urban structure. The capital city Quito was declared part of the United Nations 'human patrimony' in 1979; the UN recognition was then 'nationalized' by specific reference to the city's 'transcendental testimony to Ecuadorean culture' (INPC 1989: 37). In 1982, Cuenca was declared 'cultural patrimony of the state' to protect the colonial town core and archaeological zones. The preservation of colonial and pre-colonial architecture to represent 'authentic' Ecuadorean identity as it faces the twenty-first century is not unusual. Indeed by drawing upon the colonial past, when the relative wealth and economic centrality of Quito and Cuenca were much greater than at present, contem-porary official representations return unproblematically to a golden age of architecture (and by implication society), which disregards the explicit Spanish policy of using cities to gain control of conquered populations (AlSayyad 1992).

Among the most significant sets of monuments in Ecuador are those marking the equator line and its history. The Mitad del Mundo (Half the World) monument lying on the equator about 25 km north of Quito with its associated museums and site (Figure 3.1) is, in this respect, the most significant monument. In terms of size and number of national and inter-national visitors, it is the major national monument, being central to internal imaginings and outsiders' (tourists, travellers) perceptions. Straddling the equator line, the Mitad del Mundo complex constitutes a marker of national territory and identity, and at times is taken as a symbol of *ecuadorianidad* or Ecuadorean-ness (as in one national geography school textbook: Terán 1990).

Figure 3.1 Mitad del Mundo site, showing central tower and heroes associated with the eighteenth-century French geodesic mission

The architectural features of the Mitad del Mundo site reaffirm its national and nation-constituting status. Organized around a large central tower, which is topped by a global sphere and contains an ethnographic museum (see p. 74), the Mitad del Mundo site is set in the midst of low grassy hills. The site around the Mitad del Mundo tower, marked on its walls with the four cardinal directions, has become more elaborate and complex since its first construction in 1979 (Vera 1986). Now reached by an avenue rising slowly to the mound with the large tower, and flanked by smaller towers with busts of the various members of the first French geodesic mission, the sense of spectacle and official architecture is enhanced. Uniting symbols of science and the nation-state, the monument represents the *modernity* of the nation, a message reinforced by the masculine presence of the namers/fathers of the nation (cf. Johnson 1995). If architecture is an embodiment of national identity, then Mitad del Mundo demonstrates the significance of masculine, European-oriented nostalgia in Ecuadorean self-imaginings.

Copying another tower which had been constructed nearby in 1936, the monuments have also acted as places for the *performance* of national identity. In 1936, a national civic festival was declared by the government to mark the bicentenary of the French mission, and an elaborate inauguration programme over two weeks was devised (Vera 1986). Involving diplomats, the Instituto Geografico Militar (IGM: Geographical Military Institute), universities and historians, the inauguration united citizens in 'homage to the illustrious savants who worked on the measurement of the meridional arc, from 1736 to 1744' (quoted in Vera 1986: 54). Commemorative stamps were issued at the same time, while other similar 'pyramids' were declared national monuments in Oyambaro, Caraburo, and Tarqui, as were plaques in Quito and Cuenca (Vera 1986: 52–3). Various displays of folklore and dancing occurred in the 1970s, often coinciding with spring and autumn equinoxes.

The architectural – and implicitly national – centrality of Quito to official imaginings is replayed in the Mitad del Mundo complex on the Equator line. In a colonial style building on the site, a miniature model of Quito in the eighteenth century was created for permanent display. Representing a *pre-republican* history, the miniature contains models of key national buildings including the cathedral, presidential palace and main square, all of which continue to have national resonances in the present. The historic Quito model ties together the name of Ecuador (equator) with the centrality of Quito. However, this may be changing: other models of Guayaquil, Cuenca and Galápagos are planned. In Ecuador, monuments provide 'nodes' around which the meaningful space of the nation can be agglomerated, organized and represented to citizens (cf. Norwood and Monk 1987). Although its meanings may be contested by different groups of the elite (and others), monumental architecture can symbolize the nation and its attributes – its modernity, its scientific status, its global significance, its power and its longevity. The particular physical location of monuments in the territory can reveal the specific version of history, society and geography being promoted. The major Ecuadorean national monument ties directly to the geographical and historical relations underlying the nation's territorial configuration. By bringing together ideas of historical development and progress on the one hand, and a sense of stable continuity of the 'nation' on the other, national identity expressed through notions of time reaffirms the contemporary nation within a 'natural', historical narrative.

TERRITORY

That official discourses of nationhood have their own embeddedness in space is not as widely recognized as the historicity of national imaginings (cf. A. Smith 1986; Balibar 1990; Hobsbawm 1990; Hobsbawm and Ranger 1983; Anderson 1991). While most analysts recognize the territorial foundations of official nations, its various dimensions functioning within the circulation of signs around the 'nation' is less so. As pointed out by Escolar, Quintero and Reboratti, the emphasis is on history and the 'territory is taken for granted' (Escolar *et al.* 1994: 347). Criticizing what they refer to as the absence of a geographical perspective, Escolar *et al.* argue

> It was by means of a process of subjective representation, recognition and cartographic design, however, that the invention of the contents of a 'natural' state territory took place and that a legitimate discourse about national sovereignty was developed.
>
> (Escolar *et al.* 1994: 347).

The process of creating identification with the taken-for-granted and imagined territory is not easily summarized. On the one hand, the state institutions of schools, geographical institutes and the state-run media may create maps (both real and symbolic) of the national space, yet there are also implicit spatial representations and place-bound images which circulate in official discourses about the nation. In Ecuador, the curriculum for education on the 'natural' state territory can be examined, as too the constitution and military discourses, the literary images of Ecuador and everyday visual representations. Each of these discourses and representations interacts with the others, producing a polyphonic and complex set of official geographies for potential identification with the nation, and thereby undermining any notion of a fixed or single 'official' discourse about the national territory.

In Ecuador's constitution, a particular geographical imagination of the national territory is created, which does not accord with all social groups' imaginations (see Chapter 5). The constitution begins by saying that Ecuador is a 'unitary' state, sovereign, independent and democratic (Article 1). That the Ecuadorean state is sovereign and unitary has a specific history and a geography behind it, due to the specific context in which this nation-state emerged, as well as being a universal claim of modern nation-states. As in other countries, the faith in internal control over territory is one of the origin myths of Ecuadorean-ness; the 'origin myth' of *señorío sobre el suelo* (sovereignty over the soil) is part of the dominant class's construction of Ecuadorean national identity (Silva 1992). As Ecuador claims the subsoil for the nation, the territory is certainly represented as *sui generis*, yet being a unitary sovereign state with its own territory is not something that can be taken for granted. Since the early republican period Ecuador has been liable to territorial encroachment by Peru and Colombia, markedly in 1941 when Peruvian forces invaded a large section of Amazonian territory. The subsequent international treaty – the Rio de Janeiro Protocol of 1942 – recognized the legitimacy of Peru's action, and effectively left the Ecuadorean state with a need for new images and histories to explain its territorial transformation. However, the constitution still refers to the historical territorial claim based on the *Audiencia* (colonial administrative unit) of Quito.[5] Although certain subsequent boundary changes are indirectly referred to in the constitution, it states that 'the territory, inalienable and irreducible, comprises that of the Royal *Audiencia* of Quito with the modifications introduced by the valid treaties' (Ecuador 1993: 5). In other words, the territory is envisioned as a fundamental continuity with the colonial administrative unit, a spatial view whose consequences reverberate throughout official versions of nationhood.

As any territorial changes are seen as detrimental to the 'originary' territorial unit, the legitimacy of boundary reorganization is constantly brought into question. Official discourses on territorial nationhood continuously refer back to international boundary changes. The *historia de límites* (history of the borders) is a key component in geopolitical imaginations, not only of the military (Hepple 1991), but also of diverse officials. In interviews, military high command spokesmen repeatedly made reference to the 'problems' which result from the territorial changes. One army chief argued that the history of the borders 'is one of the most complicated and difficult problematics which face the country'.[6] While arguing that *intra*-national relations were a social contract, another military official claimed that *inter*-national treaties were bad contracts and resulted from force.[7] In this sense, the issue of frontiers and their contestation (indeed transformation) by outside ('illegitimate') powers has been locally encoded in highly forceful ideological terms which are seen to lie at the heart of national identity. According to official notions of identity, Ecuadoreans are expected to feel viscerally the injustice and damage done to the sovereign territory by foreign encroachment.

'Ecuador was, is and will be an Amazonian country' is a slogan from the 1960s circulating, not only on government headed notepaper (Whitten 1981; see also Chapter 5 in this book), but also in other fields of state action, such as education. In the geography-history curriculum, young secondary school pupils are taught about the 'discovery' of the Amazon river by one of Pizarro's officers. Thanks to the discovery and foundation of cities, school children are told, all the Amazon riverine territory is part of the territorial patrimony of the nation, recognized in the boundaries of Quito *Audiencia* in 1563 and 1740. Basing sovereignty on a historical mapping of territory, school textbooks refer explicitly to the contemporary situation, and the need for recognition of Ecuador's Amazonian rights. As one text notes, 'it is a duty of the present and future generations to demand our rights over the Amazon and its riverside territories' (García Gonzalez 1992). While these geopolitical and regional imaginative geographies lie at the root of state institutional discourses, they also influence public-oriented discourses and practices around national territory embedded in anthems and artefacts such as the flag and the naming of the country.

Marking the territory as national

According to official nationalisms, the nation is symbolized by a number of material artefacts which identify themselves as being unique to the nation or which, through their visibility and wide diffusion through the country, come to stand in for it. Officially in Ecuador there are three symbols of the nation, granted core status in the constitution's first article. These are the flag, the shield and the national anthem which function as symbols of the *patria*, (fatherland). National anthems are of generally recent origin, being commissioned or elevated to the status of anthem through decree or popular demand (Snyder 1990). In Latin America, national anthems are frequently emotional paeans to the *patria*, as in Costa Rica's 'Noble patria tu hermosa' and Chile's 'Dulce Patria'. Ecuador's current national anthem was its fourth, composed and written in 1866 with words by Juan León Mera, a major national writer.[8] The anthem extols the delight and peace now reigning in Ecuador under the 'radiant brow' of the *patria*. References to Pichincha, the province and mountain of the capital Quito are the only place-specific references in the song, but serve to map out a key place in the national imaginary. Citizens are represented in the Ecuadorean national anthem as children of the country, shedding their blood on the *patria*'s behalf in struggles to cast off the 'servile yoke' of colonialism. At the present time the national anthem, including its rousing chorus and six long verses, is printed on the back outside cover of most school exercise books, thus making it one of the most widely distributed songs in the country.[9]

Monuments to the French geodesic missions of the eighteenth century (and of 1899–1906) have been constructed in various locations in Ecuador. Lying on the equator line at 0 degrees latitude, the largest monument marks the de la Condamine mission's significance to international science and to the identity of Ecuador itself. Led by the Frenchman de la Condamine, who was funded by the Paris Academy of Sciences, the first geodesic mission in 1736–44 aimed to verify the flattening of the earth's poles. The expedition included Spanish scientists, as well as Pedro Vicente Maldonado, an Ecuadorean geographer from the town of Riobamba. Making measurements at various points in Ecuadorean territory, the mission was primarily important for generating metropolitan, but non-colonial, texts about the Spanish colonies; various accounts were distributed in French and Spanish. Knowledge about South America and the earth's properties circulated widely in Europe, due in large part to de la Condamine's writings (Pratt 1992; Brading 1991). In 1751, Condamine wrote his *Journal du voyage fait par ordre du Roi à l'Equateur* (Journal of the voyage made by order of the King to the Equator), after the Spaniards Jorge Juan and Antonio de Ulloa had published their *Relación histórica del viaje a América meridional* (Historic account of the voyage to meridional America) in 1747. As a result of the expedition and the publicity it gained in the salons of Europe, the term *equateur/ecuador/*equator entered European circuits of knowledge and representation and, mediated through them, into the post-colonial imaginings of Latin America.

Pratt has documented how Latin American creole self-fashionings drew upon the travel writings and images created by Condamine's expedition and later travellers, such as Alexander von Humboldt (Pratt 1992). New nationalist discourses in the Spanish colonies selectively drew upon their continent's representations circulating in Europe. Such transculturation of meanings and tropes, as Pratt terms it, involved a complex 'mirror-dance' of colonial meaning-making, whereby the 'science' that Humboldt produced (having transculturated American knowledges) was then reimported by creole elites anxious to give their new national identities a patina of European legitimation. In the case of Ecuador, the presence of the Condamine expedition in the territory provided (and continues to provide) a focus for creole self-identities, within the framework of a Europeanized and Euro-legitimized knowledge.

From the early nineteenth century, Ecuadorean national identity began to be constituted in relation to the Condamine scientific mission. Although still administered under Spanish colonial rule at the time of the expedition, Ecuadorean *republican* territorial imaginings and namings were to be intimately bound up with (re-imaginings of) the expedition. At the time of independence from Gran Colombia in 1830, the new republican constitution adopted the name Ecuador. During early independence when Ecuador was struggling to create a post-colonial identity, the attribution of a global 'scientific' name was appropriate, as the newly independent states wished to break into the free trade networks and Enlightenment cultures of Northern

Europe and the United States, associated with scientific progress and new civilization, compared with corporatist, hierarchical, 'medieval' and Catholic Spain. Such a realignment of national belonging in a global context was reinforced by discourses which stressed the 'civilized' nature of society in Ecuador compared with other areas on the equatorial belt. One present-day school textbook says that the scientific work was carried out in 'the *only civilized lands* whose inhabitants would not make the scientists' work difficult' (Terán 1990: 10, emphasis added).

The attribution of civilization to the country was of course for a small group of creoles only; the racialization of other equatorial regions is there, as are the racisms towards indigenous populations by the expedition members (cf. Brading 1991). Just as independent elites were 'de-colonizing' their identities, they were doing so on the basis of several pernicious racializations as they attempted to maintain 'whiteness' and racial exclusions (Pratt 1992: 141; see also Chapter 2 in this book). Overall, national identities were enhanced by foreign individuals' attribution of a name.

The naming of the new nation for an 'imaginary' line was contentious however, and continued to be so, as different groups articulated their identities in relationship to, and in dialogue with, these historic circumstances and imaginative geographies. Writers dispute the 'geographical fatality' which gave the country's name (e.g. Hidalgo 1986).

Contemporary discourses around the name Ecuador entail various racializations, but here the 'mirror-dance' of transatlantic imaginings has evolved over the past two centuries. Rather than using Ecuador as a mediated mark of whiteness, certain contemporary 'white' imaginings argue that the name Ecuador symbolizes 'blackness' in the metropole, through the name's association with Africa (cf. Jarosz 1992). Perceiving themselves in the eyes of Europe and North America to be associated with blackness and Africa, these discourses represent a value-switch in the Europeanized sign once valued for civilization/whiteness. For example, 'That it [the name Ecuador] has difficult effects in terms of international promotion is certain, because we are confused with the whole of the equator line or with the African zone.'[10]

In this context, alternative names have been suggested. Drawing upon one historic period, the name Quito represents a favourable alternative national name in some eyes. In many elite imaginings of nation, the name Ecuador displaced and erased the historic centrality of Quito. A nostalgia for an 'original', pre-colonial and thereby more 'authentic' name rather than a scientific term, becomes clear in various key texts (e.g. Terán 1990: 14). The issue of the nationality of the 'namers', the significance of the scientific work, and the geopolitics of Quito's 'national' elite, all come into play as factors influencing the debate. However, the naming of the country within transculturated European circuits of meaning and identity reinforced the creole-Europeanized groups' identity, marginalizing black and indigenous systems of knowledge. In other words, the 'marking' of the nation through naming and monuments involves contestations and contradictions.

Geography in education:
learning the place of the nation

> Geography is a very interesting science which . . . contributes to
> strengthening the spirit of the Ecuadorean nation.[11]

In many young and old Ecuadoreans' minds, the teaching of geography and
civic studies at school is associated with the teaching of 'Ecuadorean-ness'.
Forming for many years an obligatory part of the national curriculum, the
nation's territorial history comprises a key segment of the official inculcation
of national identities. The core elements include an assumed love for the land
and territory of Ecuador, a knowledge of the history of frontiers and, in many
official discourses, different descriptive tropes for the three major regions of the
country (Costa, Sierra and Oriente).

School representations of frontiers and territorial issues are widespread,
as maps are widely available and are often found on schoolroom walls. The
teaching of the 'frontier history' was an obligatory component of the curriculum in the fourth to sixth year of secondary education until 1979. Since that
time, it would appear that the materials on frontier changes are still taught
under different headings, such as civic education and social sciences. The
educational maps used for such teaching show the Rio Protocol line marking
the new border after territory was 'lost' to Peru in the 1941 conflict over
Amazonia. Nevertheless, the protocol line is represented as running *through*
the Ecuadorean map, as its legitimacy is not officially recognized. Strongly
anti-Peruvian connotations can be read into the pedagogic material used in
the Ecuadorean schools. In some texts, Peru has been labelled the 'Cain of the
South', a labelling of the 'Other' which simultaneously reaffirms the validity
of Ecuadorean identity and its territorial claims. One critical educationalist
remarked,

> The people think about themselves with stereotypes against the
> Peruvians, above all against the Peruvians, but also against other
> countries, because the vision that we get is absolutely localized [*aldeana*].
> Against the Colombians because they are thieves, and with the
> Peruvians it's because they stole our territory.[12]

In official teaching on the border issues, Ecuadorean-ness is defined as being a
nation that has lost territory (or had it stolen); being Ecuadorean means
taking the moral high ground relative to neighbouring nations. The selective
remembering and forgetting in official nationalist discourses (cf. Anderson
1991; Renan 1990) recurs in this Ecuadorean version. Although people may
not remember *why* the territory was lost to the Peruvians and Colombians,
they remember that it was lost ('stolen').

The teaching of national identity through geography is a notable feature
of certain school textbooks. In this light, the textbook, *Geografía del Ecuador*, of
Francisco Terán (in its sixteenth printing in 1993) offers a prime example. A
sanctioned component of the national geography curriculum, Terán's text has
been informing school children and the public about the country since its first
edition in 1948. Driven by a patriotic spirit, Terán believed that it was his duty

to produce a text showing the 'reality' of Ecuador in the form of a reference book to which students in the middle and higher levels of secondary education, as well as their teachers, could refer.[13] In charge of training geographical teachers, Terán occupied a highly influential position to provide geographical knowledge to citizens throughout the country. Drawing upon information from his own extensive fieldwork, the Instituto Geográfico Militar and the census, Terán published some twenty works between 1939 and 1983. His *Geografía* details the geophysical and social characteristics of the country, as well as Ecuador's position in the world. Working within a geographical tradition which highlighted local knowledges and the importance of science, Terán was concerned to present the 'facts' about the country and to enhance knowledge in order to proceed to development. To the degree that he succeeds in presenting a picture of Ecuador, its resources and the extent of their exploitation for development, Terán and other writers have much in common with the professional military geographers in state apparatus (see also Radcliffe 1996a).

Geography as a state tool

Imagining national territories in constitutional and educational geographies depends upon, and indeed refers inter-textually to, the professional mapping and cartographic represention of national space. Generally, newly independent states in the twentieth century found geography a 'necessary tool for clarifying and fostering their national identity' (Hooson 1994: 4; Dodds 1993), and Ecuador was no exception. In Latin America generally, the professionalization of geography, extensive linkages with the military and the question of territorial security have emerged in the present century particularly from the 1920s and 1930s onwards (Hepple 1991). The consolidation of geographical expertise in the hands of the state began in the 1920s in Ecuador, with the appointment of the army to the task of creating a national topographic map in August 1922 (Cortés 1960: 27). A national map, argued the Army, would permit knowledge of the patriotic frontiers and give an inventory of the country's natural wealth (including geological, hydrological, forestry, agricultural, mineral and natural resources). After gaining initial government support, the army's role was extended in 1927 with the foundation of a national map technical commission, and in 1928 with the inauguration of a six-month training programme in topography and cartography. Later in 1928, the Servicio Geográfico Militar, (Geographical Military Service), the precursor of the present Instituto Geográfico Militar (IGM), was made part of the army high command. The Servicio's work was presented in terms which stressed continuities between the 'first' Ecuadorean geographers, such as Maldonado (of the eighteenth-century Geodesic Mission) and the 'civilizing' role undertaken through geographical work. In a history, the 'pioneer' Servicio topographic team is represented as having to face the issue of ethnic difference (Cortés 1960: 29). After the Peruvian war in 1941, the Servicio Geográfico Militar was responsible for placing concrete boundary markers along the new frontier, while civilian employees were granted military designations, thereby assimilating them into the military structure (Cortés 1960).

The 1970s military takeover of government saw another significant leap in the institutionalization of geography as a cornerstone of official nationalist discourse. With a large building in central Quito, the by-now Instituto Geográfico Militar threw itself into creating information and discourses about national space. It set itself a wide remit, concerned not only with mapping and aerial photography, but also with the teaching and dissemination of geographical and historical material, folklore and anthropology (Zúrita 1960). Working with the IGM by the 1960s were several professional civilian geographers, including Terán, trained by German educational missions and in the United States, while the IGM itself was linked increasingly with inter-American organizations.

By the early 1990s, the IGM had some 500 military and civilian staff, with key positions held by military staff. Working from a modern building with a panoptic position above central Quito (Figure 3.2), the IGM was infused by a ethos of progressive scientific information-gathering placed in the hands of experts (Figure 3.3). Creating national maps for a wide audience was one fundamental role of the IGM, which prided itself on this work, arguing that maps linked them with citizens. This tutoring role began with the IGM building itself, which is decorated with two large mural maps overlooking Quito city (Figure 3.4). The maps' immediate impact and assumed recognizability were endorsed by the IGM; the head of the geographical division described the map as, 'for the public to see . . . it is one more aid, [to show them] how the borders, the territorial space, are'.[14]

The outline of Ecuador with the Rio Protocol line has become one widespread sign of the nation. Although it has not been given constitutional status as a national symbol – unlike the shield, anthem and flag – it is highly significant. By showing the Rio Protocol line, the IGM maps act to remind citizens of the 'forgotten' fact of the country's dismemberment. The IGM producers of national maps say that the Rio line is marked on maps because territorial loss is the entire country's concern, not simply a military preoccupation. That the Rio Protocol line – and not other protocol lines – is represented on the national map also serves to highlight the anti-Peruvian dimensions of nationalism, and to remind citizens of a relatively recent conflict over territory (on all protocols, see Fundación El Comercio 1993).

With the large mural maps around the IGM buildings, visitors gain an impression of Ecuador characterized by a set of distinct spatial and natural features; the choice of features and the way that they are displayed remains, of course, the decision of the army and the IGM. By including broadly defined relief features on maps, the IGM represents the three major geophysical regions of the country – the Costa, Sierra and Oriente. Compared with Terán's textbook, which treats the three major regions as distinct and relatively unconnected entities, the IGM mural map suggests that the territory of Ecuador is easily assimilable visually despite being diverse geographically. In contrast with early-twentieth-century Argentine geography that focused on harmonious interrelationships between different regions (Escolar et al. 1994), Ecuador's regions are represented as quite distinct to each other. As in Peruvian geography of the early part of the twentieth century, Ecuadorean

Figure 3.2 The Instituto Geográfico Militar, Quito

Figure 3.3 Mural inside the Instituto Geográfico Militar, Quito

Figure 3.4 Map as logo: the mural overlooking Quito from the IGM building, Quito

maps display regions with different roles to play in the country's development. In Peruvian maps and geographical knowledge, the Sierra was represented as a physical and social obstacle to development (symbolized by mountains and indians respectively), while the Amazon zone represented the future promise of development (Orlove 1993).

Similarly, in Ecuador's official geographies the Oriente is shown as the untapped resource-rich region awaiting incorporation into national development (Ramón 1990; Palomeque 1990). Ecuadorean images of its Oriente highlight the distinctiveness and 'otherness' of the region which, when incorporated into military strategic policy and national development plans, have material consequences for the populations living in that area. The 1970s military and later civilian governments saw a need for the state to occupy Amazonia, not only for security reasons but also in order to invigorate national culture and to conserve the cultural patrimony (Quintero and Silva 1991: 226). Amazonian Ecuador thus became invested with many meanings and values, through which official national ideologies were articulated.

Maps can become 'logos' for national identity in the era of print capitalism and easy reproducibility of images (Anderson 1991: 175). Certainly in Ecuador, the 'logoization' of national territory has been carried out over a significant period. In a textbook or a mural, the national territory marked with the Rio Protocol line has become an (unofficial) national symbol, calling upon shared imaginings of the nation and the truncated/mutilated nature of its space. Frontiers become highly significant places in this imaginative geography for reminding citizens of the history of national territory, and prompting identity in opposition to Peru and/or Colombia.

As well as producing maps for the public, schools and government planning departments, the Instituto Geográfico Militar is engaged in other means of representing Ecuador's place in the world to its citizens. The most visited museum and cultural centre in the country is the IGM's Centro Cultural (Cultural Centre) in Quito, which receives around 150,000 children annually, from nursery groups to secondary pupils. Together with the Mitad del Mundo complex, the IGM centre plays an important role in introducing school children (as well as the general public) to the place of Ecuador, thereby making 'a tremendous contribution to citizenship', in the words of a spokesman. The IGM centre includes a planetarium, a temporary exhibition centre displaying national and international archeological collections, and exhibits of ecology, natural sciences, history and art.

In another fundamental aspect, IGM's role has been to contribute to national defence plans. Arguing in 1960 for an institutionalized national cartography, the IGM suggested that 'the fortificatory project has to be better when the map on which it is designed is exact' (Cortés 1960: 23). The defensive aspect of the IGM's work links neatly with the national security doctrine expressed by other military institutions, such as the Instituto de Altos Estudios Nacionales (IAEN: Institute for Advanced National Studies), which all rely upon the IGM for territorial and aerial photographic information. Working under the president, the military IAEN produces many of the discourses about the nation in contemporary Ecuador. Formed in May 1972 under the government of General Rodríguez Lara, the IAEN was designed to elaborate a national security doctrine as well as 'to prepare the ruling classes (cadres) of the nation' (Quintero and Silva 1991: 223). The IAEN trained professionals in the skills necessary for sucessful development planning, and to prepare 'high level Ecuadorean professionals in the research and analysis of national reality and the international situation, to determine their influence in the security and development of the country' (IAEN 1994: 2; cf. Escobar 1995). Moreover, the IAEN training forged an organic link between the military, the state and the political class (Quintero and Silva 1991); of the IAEN graduates in 1972–7, civilians outnumbered military by five to one; in the early 1990s, the ratio was six to one.[15]

As in many Latin American countries during the Cold War, the National Security Doctrine referred to plans to deal with external threats to national sovereignty and to expand the state into the full frontiers (cf. on Brazil, Hepple 1991). Drawing upon organic metaphors and geopolitical models, Latin American national security doctrines saw a shift in emphasis from security to development during the course of the 1970s and 1980s. As expressed in Ecuador's national security doctrine in the 1970s, the military was the only competent agent for overseeing the nation's security and development, where ethnic group demands and weak unification were perceived as blocks to complete national expression (Quintero and Silva 1991: 231). Such discourses soon led to a situation where the military was seen to overlap entirely with the *patria*, linking increasingly with a discourse about the essential, natural 'national soul' of *ecuadorianidad*, independent of social groups and mobilized only via the military's development plans and discourses (Quintero and Silva 1991: 232–3; cf. Hepple 1991).

In recent years, these discourses around nationhood have changed substantially. As it oversees production of reports on the contemporary situation and appropriate planning responses, the IAEN has been central in the ideological work of adjusting official discourses on nationhood to the country's changing context. In this respect, IAEN doctrine has shifted over time becoming more concerned with issues of 'development' than security *per se*. Legitimizing its actions by reference to the 'Ecuadorean people', the IAEN focused from the late 1970s on 'national insecurities' which included – according to an Institute spokesman – extreme poverty, hunger, unemployment and racial distrust. Within this national re-presentation, the national security doctrine became not a finality but a tool to 'adequately guide the people, *el pueblo*', through which the nation could be integrated and the military represent the people.[16] In such a developmentalist discourse (Escobar 1995), nationalism is seen as something to be aimed for and achieved, through the work undertaken by IAEN; the IAEN expressly wishes to shape the national identities of its graduates. According to high IAEN officials, the Institute's aim is to reaffirm national identity: not only does this goal appear in the Institute's mission statement, but also its nickname, 'the institute of *ecuadorianidad*' (of Ecuadorean-ness), makes a joking reference to its (serious) ideological work.

As well as geopolitical and security issues, the armed forces are increasingly conscious of a role in development to be adopted in a nation no longer requiring purely military solutions (if it ever did). From the 1960s, the Ecuadorean armed forces have carried out development programmes – often in collaboration with the United States military – aimed in particular at rural and urban low-income sectors. In 1965–71, the project *Alas para la salud* (wings for health) brought free medicine to fourteen provinces, while the project *Alas para la cultura* (wings for culture) aimed to 'integrate different regions by diverse cultural means' (Quintero and Silva 1991: 220). While in government in the 1970s, the military saw the 'regional question' as the major problem preventing full national formation and argued for improved communication networks (Quintero and Silva 1991: 226). In other words, the military agenda is increasingly to produce and circulate the knowledges which are necessary for development, within their tutelary structure which puts national interest and national identity in the forefront of citizens', *el pueblo*'s, priorities.

POPULATIONS

Led by creoles who wished to break with Spanish colonial hierarchies but still retain their European roots, Latin American nations have had nearly 200 years in which to resolve the issue of who the *pueblo* are in their countries. However, the issue remains a highly contentious one, in which the primacy of national identity – as proposed by early independence leaders – has not supplanted racializations or alternative ethnic/'race' identities (see Chapter 2). Nevertheless in its ideological work of containing and explaining the country's diversity, the modern post-colonial state must represent and re-represent its population to itself and national society. This section outlines the ways in which contemporary Ecuador deals with these practices and discourses around population (see also Iturralde 1981).

In contrast to Anderson's example of certain south-east Asian countries enumerating citizens' diversity through a national census (Anderson 1991), post-colonial Ecuador has never included a 'race'/ethnic question in its national census.[17] In this respect, one dimension of Ecuadorean national identity has been the *denial* of ethnic and racial difference, although this has not prevented other arenas of the state elaborating and circulating images and 'knowledge' about the national population's makeup. In his geography textbook, Terán calculates the relative percentages of *mestizo*, white, indigenous, *mulatto* and black groups in Ecuador, focusing particularly on the Sierra and Costa (the Oriente and Galapagos are removed from the calculation because of small numbers: Terán 1990: 19). While it is not clear from Terán's text when the table was elaborated or whether it was changed in later editions, it (Table 3: 1) bears comparison with the distribution of groups given in later chapters. Making the *mestizos* and 'whites' the largest groups nationally, Terán suggested that there was a diminution of numbers of indigenous and blacks due to miscegenation and to the 'precarious economic and cultural conditions in which they live that are not favourable for their rapid increase' (Terán 1990: 20). 'White' populations are, according to Terán, increasing in numbers due to 'slow but undeniable' immigration into the country, and because *mestizos* are becoming whites, 'the mestizo groups by a selective process are losing their characteristics as such' (Terán 1990: 20). Terán's acceptance of the notion of *blanqueamiento* implicitly reproduces other official discourses on the theme of 'race' and nation. Silva argues convincingly that the official ideology of *mestizaje* is based on 'whitening', and characteristics of indigenous and black groups will be progressively replaced by 'white' ethno-cultural markers during 'development' (Silva 1991). In this respect, official ideologies of national population recognize the context-dependent and relatively fluid boundaries between ethnic-racial groups, yet officially these groups are placed in a hierarchy of value in which whiteness and moves towards whiteness are most valued.

Contradictory ideas about nature and role of miscegenation or *mestizaje* in the nation give rise to various discourses across state institutions. Military institutions, which largely draw personnel from *mestizo* middle and lower-middle class, are alert to the racial undertones to national discourses. Due to its largely *mestizo* origins, the Ecuadorean military has had much interest in rejecting racism and in rearticulating the role of non-oligarchic groups in the national project (Quintero and Silva 1991: 227).[18] The creation of a new Latin American *mestizo* race is how one IAEN spokesman saw its role in reaffirming national identity. Similarly, the army third in command argued forcefully that the armed forces 'have to be conscious of being an instrument of the nation, of the indians, the blacks, whites, *mestizos*'.[19] Claiming that there was no racism currently in the armed forces,[20] the General then referred back to the miscegenation which characterized Spanish (colonial) society, 'the mix of Arab, Jewish, Romany, and everything, because Spain has been a

Table 3.1 Terán's estimates of relative sizes of racial groups by region, Ecuador 1991

	Sierra	Coast	Average
Whites	27%	26%	26.5%
Indigenous	30%	7%	18.5%
Mestizos	42%	30%	36%
Mulattos	0.5%	28.5%	14.5%
Blacks	0.5%	8.5%	4.5%

Source: F. Terán (1991) *Geografía del Ecuador*, p. 20.

melting-pot of all these cultures'. To overcome the contradictions of *blanquea-miento* (however implicit) and ethnic-racial diversity, the military refers instead to the positive value of multi-ethnic populations. Such discourses draw upon themes of pleasant co-existence between distinct cultural traditions. An army spokesman was enthusiastic in affirming the existence of an Ecuadorean nation:

> So yes, there is an Ecuadorean nation, pluricultural, very rich in its cultural multiplicity, very rich in its ethnic diversity, . . . in 400, 500 years a nation has been structured, a nation full of defects, of injustice, which we must try to put together, and not fracture.[21]

Similarly, the IAEN suggested that 'there is one Ecuadorean nation, pluri-ethnic, pluricultural with absolute religious freedom'.[22]

The notion of rich diversity of ethnic-racial cultures is one which superseded previous *indigenista* (indigenist) rhetoric in the early twentieth century. As in Peru, Mexico and other Latin American countries with large indigenous populations, Ecuadorean intellectuals reassessed the national role of indians and, to varying degrees, reaffirmed it during the early part of the century. The 1940s in Ecuador saw the rise of *indigenista* movements, and the Instituto Indigenista Ecuatoriano (Ecuadorean Indigenist Institute, founded 1943) aimed to improve the indians through 'integration' into society (Ibarra 1992: 198). As elsewhere in the Andes, Ecuadorean *indigenistas* raised the nationalist standard on the glories of *pre*-Colombian civilizations, strongly separating them from the present-day indigenous 'unfortunates'. Terán's textbook for example emphasizes the value of pre-Colombian indian cultures, which he calls the 'primitive trunk of nationality' (Terán 1990: 17).

Indigenista intellectual and political projects did however change during the decades after the founding of the Instituto Indigenista Ecuatoriano. During the 1950s and 1960s, education and especially literacy training was seen as the most efficient means of accelerating integration, while the term 'indigenous' was dropped in favour of 'peasantry' (as in Peru in the late 1960s). A new strand of *indigenista* discourse emerged during the 1980s as Amerindian organizations and the election of progressive governments permitted a reassessment of previous models. *Neo-indigenista* (new indigenism) discourses emerged and continue to circulate, presenting a critique of traditional *indigenismo* (Ibarra 1992: 175). Rather than emphasize integration, new indigenism stressed the autonomy and validity of multiple indigenous cultures, and the importance of respect for difference. Drawing upon ideas of participation in decision-making and 'integrated rural development', governments proposed the recognition of indigenous organizations as 'privileged interlocutors' and respect for the multi-cultural nation. For example, the Ecuadorean 1980–4 National Development Plan aspired to the 'protection, conservation and investigation of vernacular cultures', in which 'acculturation of their members does not imply renouncing their own cultural identities' (Ibarra 1992: 207; see also Berdichensky 1986).

Rather than having ethnicity forming an unproblematic template for national identities, in Latin America the relationship between indigenous ethnicity and the nation is still highly problematic, and resolved differently in different societies. In a multi-ethnic context, Mexican nationalism claims direct descent from pre-Colombian populations (N. Gutierrez 1990; Brading 1991). Such a move to keep indigenous culture in the past may entail the distancing of ethnic issues from the contemporary nation and the denial of contemporary indigenous demands. While not as *indigenista* in its rhetoric as Mexico, Ecuador lays claim to its population's rich diversity at the same time as avoiding the land distribution question, which is so central to indigenous identities and politics. The stress on mutual recognition of cultures in official discourse also remains compatible with the idea of a unitary nation, a key element of administrative concerns especially among the military.

Disciplining the national social body

At the same time as being concerned to enumerate and classify citizens, the modern nation-state is increasingly engaged in embodying national values in its citizens. The national social body can thus been seen as subject to the disciplinary effects of state discourses and practices, which attempt to inculcate national identities and behaviours. In Ecuador, such disciplinary practices include military service, and military training of school children.

As the territorial conflict with Peru waned in national agendas (at least until the border conflict in January–February 1995), the military has been reworking its discourses to claim a long-term interest in national development and peace (Molina Flores 1994). A far cry from the 1970s Argentine military response to the 'enemy within', the Ecuadorean military is engaged in practices to form citizens who can carry out its (declared) agenda of peace and development. These interests are expressed in discourses and practices which lead to embodiment of citizenship, and what is termed the 'treatment' of the nation's body. In contemporary military discourse, the country is seen as a human body which needs appropriate 'treatment' in order to guarantee peace and order; in this respect, the use of the organic metaphor by the Latin American military has taken a further twist in its long history (cf. Hepple 1991). The indigenous uprising in 1990 was seen as a crucial moment for the definition of this construction.[23] According to one key ideologue in the armed forces, 'the uprising [*alzamiento*] made us think . . . I think that preventive medicine in the human body as in the social body, is important.'[24]

Accordingly, the military began development work in various regions – particularly in the Andean province of Chimborazo, 'the province with the highest proportion of indigenous', said the same military spokesman (also, not incidentally, a major site in the uprising), and in the Oriente. In every project, the expressed military aim was to work alongside indigenous communities, rather than carry out acts of paternalism. In each case, the treatment of the *bodies* of citizens is a recurrent theme: other dimensions of 'development' which do not deal directly with intervening physically in human bodily development are relegated to lesser importance. In Chimborazo for example, the main projects undertaken included 'health, sanitation, loaning our doctors, campaigns of "de-parisitization"', and in Guayaquil, 'permanent work of vaccination, and control of illness such as epidemics of dengue fever, rabies [*sic*]'. Other, discursively minor, elements include building of roads, bridges and provision of drinking water.

Another dimension of the disciplining and formation of national citizens was attempted in the armed forces, voluntary, school-based, military instruction programme which aimed to create 'morally, civically and physically apt citizens to confront the harsh reality of our country' (Molina 1994: 141). Drawing in school children from a wide range of educational centres, the programme ran on Saturday mornings and gave children basic instruction in firearm use and control, as well as development-type work in forestation, first aid and civic education. The programme was designed to bring together students of different ethnic and class backgrounds, such that they made friends outside their usual social circles. Open to both boys and girls, the elementary military training appears to have an agenda of creating 'self-surveillance' among future national citizens. The moral education involved in the programme implied, a military spokesman said, that the school children 'know that the next day [Saturday] they are going to have an intense workload, physically and intellectually, so they take care not to use drugs, alcohol'.[25]

The process of embodying national moral values in school children appears to be behind the new Saturday training programmes, which involved up to 9,000 students in Guayaquil alone in 1994. As in military service and the IAEN programmes, the military school activities have an explicit national goal – that of increasing levels of nationalism among citizens – and an implicit goal of directing human energies to the activities and behaviours that the military believe are most conducive to national development. The school curriculum has elements designed to lead in the same direction, inculcating national values and identification in primary and secondary pupils. However, the key ideologues and practices found in education are distinct to those in the armed forces, highlighting the diversity and polyphony of official discourses of nationhood.

Schooling national citizens

According to the 1990 census, around three-quarters of the population have primary education: just over 50 per cent (50.9 per cent) have some primary education and a further quarter have some secondary education (25.9 per cent).[26] Clearly, differential participation rates in education by location, gender or age, may affect takeup of the curriculum. However, the factor of age would appear to be the only significant one at the present time; in the 1990 census, lack of schooling was more notable among older generations, although in 1989 an alphabetization campaign drew in older learners.[27] As more rural schools were founded and staffed, the rural–urban differential in education lessened. Men and women have similar rates of participation and completion in education in all age groups except the over-65 years cohort (Ecuador 1991).

Within the curriculum currently being developed, the education model comprises three relevant components – language, civics and extra-curricular (though ubiquitous) ceremonies enacted in schools. Language is a key nationalist issue, indicating through its use the nature and extent of the imagined community with which the individual user communicates. Although the Spanish and Portuguese languages were imposed in Latin America, the fate of colonial languages after independence varied considerably, with Portuguese in Brazil for example being adopted unproblematically in the post-colonial period. By contrast, Paraguayan anti-colonial leaders resurrected the 'authentic national (pre-colonial)' language of Guaraní as official language after independence from Spain (Urban 1991). Neither Brazil or Paraguay is 'more' indian than Peru or Ecuador, yet in these latter countries the persistence of large indigenous populations has not led to the official adoption of Quechua/Quichua as a state language. In Ecuador, Spanish is the official language while indigenous languages ('aboriginal languages') are seen as 'part of national culture'. Estimates of Quichua-speaking vary from 9.4 per cent to 15.2 per cent of Ecuador's population (Knapp 1991), while other indigenous languages are less widespread. Ecuador's official emphasis has always been on universalizing Spanish and the non-universalization of Quichua and indigenous languages. The recent introduction of a bilingual education programme in the country has not challenged this basic attitude.[28]

Historically, the civic education component of the curriculum has been where official national identities are primarily forged. Based on moralizing the young, civic education is characterized by an exhortation to love and respect the nation and its symbols. The *pre*-school curriculum for example includes a section on 'our *patria*' where the children learn the country's name as well as to respect the *símbolos patrios* (patriotic symbols) of the flag, anthem and shield. Among various 'corners' for activities, it is suggested that a 'patriotic corner' can be made with the children (Ecuador 1992). As well as learning to recognize symbols, all school children are expected to *behave* in particular ways which reaffirm their position as future national citizens. By the second grade of primary, children are expected to know how to behave in front of the patriotic symbols, to know about the resources and pluriculturality of the country, and to know their 'duty to respect and defend it' (Ecuador 1992: 150–62). Education in subsequent years refines on these basic precepts and knowleges (García Gonzalez 1992).

One (militarized) ceremony in schools is the 'swearing before the flag', by which students swear allegiance to the fatherland (*patria*) in order to graduate from primary and secondary school.[29] In a brief oath, pupils swear loyalty to the national state, the flag and the nation. Although school children laugh at these civic ceremonies, they still arguably take a romantic view of the *patria*.

Among indigenous populations with minimal sustained contact with the market and the state, the disciplinary aspects of education are highly visible. Among the indigenous group of Huaorani in the Oriente, state education is as much about changing bodily habits and clothing as it is about learning the curriculum. Schooling brings 'civilization', that is a new set of behaviours associated with the modern status of citizenship: 'when [Huaorani school children] go outside [to the national public sphere], they must be washed, wear clean clothes, show self-restraint, and avoid bodily contacts' (Rival 1994: 16). The creation of citizens whose public behaviour and deportment are suitable and 'national' is arguably part of the implicit agenda of state educational establishments.

Certainly, the ways in which the patriotic symbols are used in ceremonies emphasize the primary significance of national identity within the education system. Critics also highlight the militarized nature of civic education in Ecuador, pointing to the flag-swearing ceremony in schools, as well as use of the flag and maps with the Rio Protocol. The commemorations of geopolitical dates in the civic moment and history return endlessly to the trope of territorial loss, and the existence of an imagined community around these features. The 'happenings' remembered in civic education are key historic moments in Ecuador as well as key geographic points, together providing a narrative of national time and national space.

Displaying the nation: museums and society

The role of museums in the presentation of national societies and their histories has been widely recognized (e.g. Kaplan 1994). As containers for 'national' objects and as presenters of populations and pasts, museums serve a multiplicity of nation-building functions, most related to the deeply political goals of the 'museumizing imagination' (Anderson 1991: 181; Sherman and Rogoff 1994). Moreover, national populations and social relations are represented and 'explained' in national museums in ways which enhance processes of 'imagining' co-citizens. By delimiting where and how different groups live in the territory, museums present the 'facts' of co-existence of national lifestyles, at the same time as they picture (or silence) groups and thereby play a crucial role in re-presenting national society to itself, especially in ethnographic museums where the subject is national cultures (Kaplan 1994). In representing cultural difference between national groups, the unifying and 'nationalizing' of citizens can become problematic, given the emphasis on difference rather than sameness. Museums can thereby reinforce hierarchies of difference, while at the same time celebrating diversity and the existence of 'colourful' minority cultures. The problematic of difference/ nationhood is one with which Ecuadorean museums, both state and private, have to deal, revealing a pattern of representing diversity within an overall discourse of nation-building. Distinct ethnic groups are represented as within, and as part of, the national community, their diversity enhancing the richness of the nation-state and not threatening it.

The ethnographic museum at the Mitad del Mundo site is one of the most visited in the country. Run by the Provincial Council of Pichincha (which includes Quito city) since its transfer from the Banco Central del Ecuador (Central Bank of Ecuador), the museum is visited mostly by school children, who – especially if they are based in Quito – probably visit the museum twice by the time they finish primary school.[30] The museum is arranged over several floors in the central monument, and visitors move down from the top floor past displays of different cultural groups. Each display contains a name plaque, a map of Ecuador showing in red the group's distribution, some artefacts and several large-format photos on the walls behind the artefacts. Visual information on each group is reinforced by short explanatory panels, giving details of lifestyle, origin of names and current population numbers.

The impression given by the ethnographic displays is of multiple ethnicities, each with its own place in the nation and its distinct material culture. Indigenous groups provide the bulk of these ethnicities, although black groups and *cholos* (indigenous-*mestizo* groups) are also included. The glaring absence is of whites or *blanco-mestizos*, who have no section devoted to them, although they appear as individuals in one section's photos. As in other parts of the world, whiteness is invisible and unquestioned in Ecuador's national ethnographic museum (cf. Ware 1992); neither are Lebanese and other Mediterranean immigrants represented. In other words, the museum talks about different 'human groups', within which racializations of indigenous and black populations occur. Black populations are associated particularly with music and dancing, a pattern noted widely elsewhere in the north and in post-colonial societies (Jackson 1989). For example, the display on coastal black populations contains wooden musical instruments, and photos of men and women dancing and playing music. The Sierra Negros are said to have the same culture as coastal blacks, and their presence in the highland Chota valley is attributed to slavery.

Diverse 'human groups' are clearly located in specific areas in the country via maps. Different cultures are mapped as 'grounded' in particular and un-overlapping locations. Population and place are represented as equivalences; places become defined by populations, discursively linked with bounded, subnational places. Maps purportedly show the original locations of each group before migration, although no mention of this movement means visitors have to make their own interpretation. The predominant narrative is of unmixed cultural traits and racial characteristics, and that distinct groups live separately in different parts of the country. For example, the *montuvios* are labelled 'a *mestizo* peasant group of the Ecuadorean coastal plain' with no link made with indigenous or *mestizo* peasants elsewhere. Indigenous groups in the Oriente are linked explicitly with geographical features, along the major Oriente rivers. The direct relationship between one place and one culture cannot be applied in the museum section on the Sierra, where groups are instead represented in relation to provinces. For example, the 'groups of [the province of] Imbambura' are described as 'distinct human groups, among them some indigenous, such as those of Otavalo, Illumán, Peguche'. The presence of white-*mestizo* groups in Pichincha province (located mostly in the capital Quito) is glossed in the display as 'a significant influence of urban culture'. The display then reasserts non-white identity: 'there exist groups who keep alive their tradition, whether indigenous or *mestizo*'. In this section, unlabelled white-*mestizos* appear in a photo-mosaic of individual faces, with cityscapes of Quito behind. The processes of miscegenation and rural–urban migration are unsettling to the idea of distinct neat cultural groups. The museum treatment of the mixing of urban and indigenous cultures, called *choloficación* in the Andean countries, illustrates this (Abercrombie 1991). The designation *cholo* is glossed in the Ecuadorean museum as 'a general term in the country to designate a person born from indo-hispanic mixing [*mestizaje*], in whose behaviour the indigenous world predominates' (cf. Bourricaud 1975). Nevertheless, the widespread fact of the process is sidestepped, and the implications for 'national' culture avoided. Rather than attribute *cholos* to a general national process, the phenomenon is highly localized and gendered in the Mitad del Mundo museum. Being a *cholo/a* is linked with Cuenca, the largest southern Sierran city, and with women, rather than being represented as a widespread process involving women and men around the nation. In the museum, two life-sized female models of *cholas cuencanas*, or female *cholas*, from Cuenca display the clothing associated with this group.

The activities displayed in the museum emphasize markedly the Catholic and secular festivals, such as bull fights and firecrackers. In emphasizing cultural difference in terms of these customs and traditions – and by down-playing work, domestic life and political activities – the museum re-presents a picture of parallel folk traditions.[31] Human groups, some of them racialized, are drawn together in the national space, metaphorically and literally through the prominent map in front of each display. The use of the map-logo throughout, and displays of the environmental and human resources of the country, reinforce this message.[32] The culmination of the play between heterogeneity and national unity come together in the slogan shown on the last wall before the exit: 'Nature must be preserved and the national identity strengthened. Only then will Ecuador realize its historic destiny.' The museum's discourse presents ethnic-racial differences as part of another *inventory* of the country, as an (invented) 'parallel series' of national compo-nents through which the nation is organized and imagined (cf. Anderson 1991: 169).

Organizing the displays of national society

In Ecuador, the institution which oversees museums and the protection of cultural and historical works is the Casa de Cultura, literally the House of Culture. Naming the institution a *House* of Culture resonates with ideas about nations being 'domestic genealogies' (McClintock 1993a). The Casa de Cultura becomes the national/'domestic' institution for cultural artefacts belonging to national 'family'. Indeed, the Casa de Cultura continues these familial tropes in its publicity: in posters, the Casa de Cultura building in central Quito was called 'our house'. In the national house, there are those who are expected to provide guidance (the fatherland) and others who learn and follow (the people), in a democratizing process. Visitors to the Ecuadorean Casa de Cultura are reminded that 'culture is the most important and significant democratic act of a people'. Although referring to democracy, the official conception of culture tends to underplay the elite criteria of what is defined as the national (familiar) patrimony. The main Benjamín Carrión museum of the Quito Casa de Cultura draws a wide audience to displays of national culture, particularly represented in oil paintings of the nineteenth and twentieth centuries. As well as foreign and national tourists, a large number of school children visit this major national collection. Paintings include portraits of the 'liberation hero' Simon Bolívar, past presidents of the republic, scenes showing events or places of 'national' importance, landscapes and portraits of individuals (see Chapter 5). Black and indigenous figures are not well represented in the museum collection, although a few unattributed naive paintings have unnamed dark *mestizo* and indigenous women and children as subjects ('*Chola* with child' is a typical title). However, the experience of the museum is informed by the playing of 'Andean' pipe music over the loudspeakers.

Working under the Casa de Cultura is the Instituto Nacional de Patrimonio Cultural (INPC), specifically charged with the preservation of historical cultural goods. Similar to the cultural institutions described by Anderson in late colonial South-east Asia, the Ecuadorean INPC also has Latin American parallels such as the Instituto do Patrimônio Histórico e Artistico Nacional (IPHAN: National Historic and Artistic Patrimony Institute) in Brazil and the Instituto Nacional de Antropología e Historia (INAH: National Institute of Anthropology and History) in Mexico (Dickenson 1994; Morales-Moreno 1994). The Ecuadorean INPC is responsible for the conservation, preservation and rescue of the cultural artefacts of the nation, particularly of monuments, art and archaeological work.[33] Drawing upon nationalist discourses embedded in the Artistic Patrimony Law of 1945, the law outlining the role of the INPC declares 'the State must conserve the cultural patrimony of the people, as a baseline for their nationality . . . in accordance with their traditional ways of life and ancestral customs to the present' (INPC 1989: 7) Founded in 1979, the directorate of the Institute includes military, church, government and education representatives. As well as conserving and protecting the national heritage, the INPC is charged with creating an inventory of patrimonial holdings in public and private ownership. As a form of ordering, the material culture inventory acts as a similar administrative device to the census (inventory of populations) and maps (inventory of territory). The INPC inventory aims to identify all 'national' relevant monuments and pieces of art and then declare them as 'national patrimony'. Once objects have been declared national patrimony, regulations control their movements (especially out of the country), restoration, exhibition and research on them (INPC 1989: 8). All citizens in addition to the armed forces, civil police and customs officials are obliged to 'collaborate in the defense and conservation of the Ecuadorean cultural patrimony' (INPC 1989: 8). Protection of 'national' culture is thereby embodied in citizens, whose national belonging is defined both by a legal obligation and by the always-already assumed direct relationship between citizens and 'their' artefacts. In this way the origin of the artefacts in pre-colonial or colonial periods and often from elite groups is glossed over, such that citizens have immediate, present-day and direct (although largely hypothetical) 'ownership' of the cultural patrimony.

Citizens making museums

In addition to the traditional format of museums, the Ecuadorean state attempted to involve its own citizens in creating and provisioning museums in the early 1990s. However, the impetus for new museums came from the state, carrying with it an implicit agenda for the control of (indigenous) populations and artefacts. The idea of creating community museums gave rise to the construction of museum buildings alongside archaeological sites and indigenous settlements in the Sierra and Oriente. In the course of creating new museums, INPC staff went to villages to persuade local people to hand over archaeological finds, which previously had been sold to national or international collectors. After creating an inventory of the 'donated' pieces, the INPC placed them in local museums. The INPC also encouraged the local population to guard and 'museumize' any other archaeological objects they found. Strongly condemning sales, the INPC was intolerant of what it perceived as peasant irresponsibility (displayed in drunkenness) and it constituted itself discursively as a guardian institution overlooking peasant museums.

One such local museum was created in Cochasquí, a recently discovered pre-Inca site in the Quito area, where the local indigenous organization collaborated with the INPC archaeological and museum staff in the creation of a local museum. Built to include houses, implements and goods from contemporary communities nearby as well as archaeological finds, the site offered a showcase for the new linkages between national history and contemporary national society.[34] The local peasant Unión de Comunidades Campesinas de Cochasquí Moncayo (UCOPEN: Cochasquí-Moncayo Union of Peasant Communities) brought together some forty villages to build the museum, a process which led to the 'rethinking and revaluing' of their cultures. These sites also provided a place for the elaboration of new, 'invented' ceremonies rearticulating local and national identities: Cochasquí saw various folkloric musical events and a solstice-harvest dance and music festival soon after its inauguration.

While the multiplicity of cultural groups has been acknowledged by the Ecuadorean state in its museums, the question remains of how this diversity is explained within the idea of a unitary nation. Looking at Latin America, Urban and Sherzer (1991: 12) argue that the national state exhibits two contradictory tendencies; the state is attracted to difference and the exotic, providing uniqueness and distinction to society, yet this difference potentially threatens the sovereign jurisdiction of the state. The Mitad del Mundo museum illustrates both of these tendencies: it flaunts the multi-cultural dimensions of Ecuadorean society, while avoiding reference to power differentials between groups. The peasant museum at Cochasquí proudly displays indigenous artefacts (from past and present) under the tutelage of state insitutions. Whether national citizens identify as belonging to a racial-ethnic group, the official discourse around population allocates a person to only *one* subgroup running in parallel with others. As well as belonging in only one ethnic group, populations are represented as belonging to *one* geographical location – identities are assumed to be constituted in direct, unmediated and singular relationship with one place.

By outlining the major themes of Ecuador's official nationalisms, this chapter has described and analysed the discourses, practices and representations about nationhood which the state thinks citizens should relate with. Official discourses of national identity are predicated upon a belief that the state and its institutions can intervene positively in citizens' acquisition of a national identity. Taking an active role, the state creates (and re-creates) texts, maps, inventories and representations of the country that citizens experience during the course of their lives. Official discourses and practices of nationhood can thus be defined as explicit and deeply embedded within the state. The state-initiated nationalisms attempt to set boundaries, boundaries which are both material and representational, and which act to set clear differences between the national on the one hand, and the local and the international on the other. The state's agenda of order and discipline, and the constant efforts to clarify social and spatial boundaries between 'us' and 'them', 'here' and 'there', are found in a variety of spheres. Yet the will to order and clarifications sought by the state in its nationalisms are constantly contested and the boundaries transgressed. In the practice of official nationalisms, the discourses and practices are not uniform or entirely consistent. Various conservative, racist and Quito-centric characteristics of official ideologies run alongside discourses about human rights and the peaceful co-existence of diverse cultures and ethnicities, even within military institutions. While history is presented as a narrative of battles over territories and geopolitical military concerns, the role that the military has carved out for itself is now more concerned with development and the moral welfare of citizens. The map of the country – marked indelibly with the Rio Protocol line – is ubiquitous in the country, yet the logoization of this symbol also resonates with notions of Amazonian promise and the compactness of the country. In other words, the official discourses of nationalism are in flux at any one time, and face contestation around certain key issues by groups influential in the state.

In talking of the relationship between myth, history and identity, Friedman argues that the temporal continuity of identity is established (or re-established) by means of spatial discontinuity (Friedman 1992: 194). In the case of Ecuador's official discourses of nationhood, the permanence of national identity is certainly predicated upon a (material *and* highly symbolic) spatial discontinuity with other neighbouring countries, particularly Peru. Furthermore, this national geography is reinforced by the tracing of national 'origin' to the invented Kingdom of Quitu, reiterating the idea of a long-term national space abutting Other nations. By calling upon frontiers as signs for national difference, official discourses of nationhood re-represent the nation as historically continuous and geographically discontinuous, by interweaving space and time.

Moreover, official national discourses in Ecuador are characterized by a prominent geographical imagination. The implicit, and at times explicit, attribution of populations and different livelihood activities to different regions or places within the nation acts to create an imaginative geography of the nation. Although diverse and at times contradictory, it is arguable that official discourses rely for their effect on an implicit organization of (national and non-national) places into a *hierarchy* of value, in which certain places in the nation are less valued than others. No less important are the official discursive effects which spatialize populations, whereby certain areas are racialized in various ways and racial groups seen to be located in particular areas. The presentation of Quito as the centre and epitome of nationhood recycles notions of Andean civilization and superiority (based on whiteness, Catholicism, elite landowning families), an example to nationals elsewhere in the country. The implicit imaginative geography of official national identities is an often unrecognized yet crucial dimension in the analysis of discursive formations of state and official ideologies of nationhood. The degree to which Ecuadorean citizens actually take on board these discourses of hierarchy, population, history and territory – and in what ways they recirculate and reconvert them – are the subjects of the following chapters.

4

CREATING BELONGING

Cultural formations, identities and correlative imaginaries

Earlier discussions in this book provided a foreground to the complexities of the theorization and lived expressions of national identities. The fluid, contradictory nature of identities reinforce an understanding of the importance of the sites within which identities are constructed and 'structures of feeling' experienced. Chapter 3 examined the ways in which an official discourse of the nation and thereby national identities is constructed in relation to one nation-state, Ecuador. Clearly, there are important historical markers for the articulation of 'the nation', juxtaposed to the important territorial claims (which have focused on the Ecuador–Peru border dispute) which are part of the school history syllabus coming to occupy a special place in an individual's understanding of herself or himself as Ecuadorean. Thus, time and space are both crucial to the configuration which comes to be known as a national identity. But historicized time and geographically located space are not simply axes around which identities and nation-states are built. As Chapter 3 makes clear, both are re-presented to the populace within the public realm via schooling, national monuments and museums as part of the process Althusser (1972) called interpellation, a process which 'calls forth' subjects, drawing them into a discourse which itself shifts the sense of identity. Clearly, the official discourses of the nation are organized to do this from schooling and through public ceremonies, but it is not simply top-down ideological work. Children in Ecuador visiting the Mitad del Mundo museum showed evident delight in the 'exhibits' and, we want to suggest, in the ways in which they could place themselves as part of the plurality of national ethnicities and cultures 'exhibited' for their education and induction. The children are not, therefore, simply passive recipients of messages but active interpreters of a world re-presented in the museum, fashioned from the lived experience of Ecuador. In placing themselves in the nation as subjects called forth by the exhibits and the space of the museum, the children were practising a correlative imaginary which set the individual and the nation side by side, one and one, 1 and 1. Thus the interpellation and the structure of feeling of a national identity within a specific nation could be sustained via the inclusivity signified in this case by the 'exhibits' which generated such a positive and active response from the children. These processes are one part of the national story which promotes inclusion and horizontal integration consistently undermined, in effect, by the exclusions of region, class and racism.

This raises the concerns of this chapter; having tried to elaborate an 'official' discourse on the nation we are immediately faced with the binary opposition, the 'popular' and its place in the construction and re-construction of national identities. But it will also be clear that part of the power of the 'official' derives from the ways in which the national story is woven together using elements from the popular. Thus, alongside other writers, most notably Rowe and Schelling (1991), we would argue against the separation and the binary 'official'/'popular', using instead the insights from Gramsci (1971) that ideological hegemony is powerful in part because it is unstable and it brings together the dominant and the subordinate. Foregrounding this instability, Winant (1994) has more recently argued for the notion of a 'decentred hegemony', combining the work of Gramsci with insights from post-structuralist accounts of power, the social and the subject. In their illuminating and elegant book Rowe and Schelling argue that the problem this raises is where to locate the popular. They suggest instead that it is constantly in process within specific sites that construct and reconstruct popular culture and subjects in terms of the past and the present. Equally, they hold fast to the notion, as we do, that popular cultural forms are power relations – the most important insight from Antonio Gramsci – culture and its moment are political. In adopting this view we overcome the problematic binary suggested in some writing on popular and elite cultures, most especially developed in the work of Gellner (1994), for example, in relation to nationalism. The paucity of Gellner's analysis and concentration on induction into high culture is completely undermined by the phenomenon of football in Latin America, to use one example explored later in this chapter.

Football is clearly part of 'modern' popular culture delivered and consumed through the mass media forms that are global and local simultaneously. As Rowe and Schelling (1991: 8) note, 'Modernity arrived with television rather than the Enlightenment', which may be an overstatement but it does mark the importance of mass media forms in the Latin American context. It is based upon an encounter not a passive reception, rather an engagement or 'mediation' which itself generates and sustains differential receptions so clearly seen in relation to televised football games.

Mass communication is one part of the story of popular cultures but folkloric elements are also important and again show the complexity of the articulation betwen the official and the popular (see pp. 74–5 on the ethnographic museum). Again, as Rowe and Schelling point out, 'it is probably valid to say that in Latin America the idea of folklore is bound up with the idea of national identity, and has been used by the state, among other things, in order to bring about national unity' (1991: 4). This provides an interesting contradiction in which difference is made malleable in relation to official accounts of the nation and transformed into an under-theorized multiculturalism in support of a 'fictive ethnicity' as the basis of national unity. Not surprisingly, it has encountered some problems: Ecuadorean postcards favour photographs of indigenous peoples, making of them folkloric artefacts for consumption by the tourist industry. More recently in Ecuador there has also been an attempt to rescue an indigenous past that is specific to Ecuador,

not shared with Peru or Colombia. Thus, at the Instituto Nacional de Patrimonio Cultural (INPC) we heard news of new archaeological finds that
would rival those in Mexico in terms of the pre-Colombian past. This invention
of tradition is assumed as the responsibility of the middle and upper classes
who fashion a past that was indigenous but which is separated from contemporary indigenous peoples and cultures. Omitted from this are the mediations
between the popular and the official expressed in relation to Spanish
colonization of the indigenous peoples:

> the spectacular victory of Spanish colonization over the indigenous
> cultures was diverted from its intended aims by the use made of it. . . .
> the Indians [*sic*] often used the laws, practices and representations that
> were imposed on them by force or by fascination to ends other than
> those of the conquerors. . . . They metaphorized the dominant order:
> they made it function in another register.
>
> (de Certeau 1984: 31–2)

This cultural reconversion and the continued power of local codes of reception
was, and is, most transparent in relation to Roman Catholicism, a terrain
discussed later in this chapter.

Cultural reconversion is also to be seen in the attention to artefacts now
generated for the tourist trade which ensure that village economies are
integrated into the capitalist development process through the sphere of
circulation. Addressing this issue in relation to craftsworkers in Mexico,
Canclini (1993) points to the subtleties of capitalism in which so-called traditional crafts are not destroyed in the wake of modernization but incorporated
within a culture of consumption that generates a multiplicity of markets,
not only mass markets as part of the post-Fordist era. Combining a Marxist
economic analysis with Bourdieu's (1977) notion of cultural capital, he
suggests that crafts as cultural capital have, in part, been able to resist defor-
mation and obliteration in the processes of global capitalist development.
Through the production and circulation of these craft-based commodities
there is a base for resistance and a collective subject able to reorganize
the terms on which artisans enter the market. This base would, he suggests,
provide an alternative production of culture and contribute towards
an 'authentic' popular culture which seeks in Gramsci's sense to alter the
commonsense. Canclini's work, suggests Yúdice (1992: 20), provides an
analysis which suggests the multiplicities of modernity in Latin America as 'a
series of necessarily unfinished projects'.

Our intention in this chapter is not to catalogue the infinite variety of
popular cultures, but to consider the ways in which popular cultures as
politicized moments contribute to and contest conceptions of national
identities. Importantly, moments of popular culture are moments of belonging and this is one very powerful way in which the seduction of national
identities takes place offering a space, a 'home' to individuals within the
broadest time/space frame.

NATIONAL TIME/SPACE

Foucault in an early paper called the modern period 'the epoch of space' and suggested that 'The anxiety of our era has fundamentally to do with space rather than time' (1964). The separation between time and space suggested in Foucault's remarks has been both reproduced and contested by later writers; the famous post-modern account of time/space compression from Harvey (1990) and others belies the distinction. Yet, as Doreen Massey (1994) suggests, although 'space' and 'time' are constantly invoked there has been little work of serious conceptualization. Massey criticizes Laclau (1990) for depoliticizing space and offers, instead, a notion of space articulated with the social and constructed by social relations which converge on specific sites. Thus, space, power and identity are all three constituted together and re-constituted constantly, what Massey (1995) calls place-identity. Place-identity would seem to naturalize 'place' rather than 'space', one source of Massey's critique of Laclau and Mouffe. Consequently, space is not privileged in the definition of identities but through the web of power relations identities are constantly made and re-made. Clearly, these relations and their imaginaries are circulated and reconverted within time. Time and temporality are multi-dimensional – capitalist time invented the notion of 'free'-time (Adam 1994); Foucault's (1977) disciplinary modes of modernity incorporated workers as the proletariat organized, body and soul, within new conceptions of time related to capitalist production imperatives. Such temporality is quite separate from the rhythms of cyclical time located with seasons and festivals and yet a degree of certainty re-emerges precisely within official accounts of national time through commemorative dates. Cyclical time was ruptured by capitalist development but not obliterated. It exists and is reproduced through festivals, religious calenders and the rhythmn of rural production, contributing to the way that the nation is reiterated. Time and space are articulated with the national to bring into focus an account of the nation and national identities in relation to specific histories, into which individuals can insert personal biographies and thereby connect with a sense of national belonging. Memory and acts of re-membering within spatial/territorial boundaries are crucially productive of national identities and these productions are revisioned within popular cultures through festivals, television, songs and stories articulated with current politics to re-cast the political imaginary.

These preliminary remarks are an introduction to what we have called 'national time/space' as a way of foregrounding the role of popular culture in generating shared conceptions of histories and cultural and geographical boundaries. Popular culture contributes to the ways in which individuals are able to place their biographies within the larger stage of the nation-state producing a correlative imaginary which is so important in the generation of national identities. Clearly, there is no one version of this process, but a multiplicity of stories that have diverse cultural and regional roots; some are refashioned for popular consumption at a general level. The Mitad del Mundo is one example of a state-sponsored writing of time and space. Equally, but in counter-hegemonic mode, the '500 years of resistance' campaign reversed the official history and changed the vision of America for those in Europe and North America as well as at home. The campaign also, most importantly, reversed the notion from Europe of 'our time' against 'other time' – the Otherization of time belonging spatially outside Europe or in the pre-modern era, both coming together in the Eurocentric view of the Conquest (Fabian 1983). It is from this complexity that the multiple sense of cyclical and 'modern' national time is generated.

One interesting example of this in Ecuador was a very popular 1994 television programme called *La T.V.* produced and presented by Freddy Ehlers, a white, elite Ecuadorean with a metropolitan background. *La T.V.*, watched by 3.8 million viewers in May 1994, was a current affairs, documentary-style programme which sought to bring issues as diverse as oil pollution in Amazonia and current political scandal together (not unlike the Peruvian *24 Hours* programme).The programme was one way in which the media tried to define within Ecuador a contemporary 'agenda of concern'. In discussions with Freddy Ehlers, it became clear that the programme not only sought ways to define a national present, but also was seriously concerned with a revisioned past. Re-presenting national time no longer began with the Spanish but was now constructed to include indigenous peoples who, it was claimed, were distinctly Ecuadorean and provided the basis for Ecaudorean national identity. Thus, the formerly excluded indigenous peoples were now to be included in a new hegemonic project around an Ecuadorean nation in which the middle and upper classes could exercise a form of patronage ('our traditions') and become part of the claim for an authentic Ecuadorean past and present, in a move parallel to that of education and the INPC (discussed in Chapter 3). It was also important that this re-visioned Ecuador was to be placed within a re-imagined Latin America more generally, itself claiming a distinctiveness from American cultural imperialism. This is a distinctiveness fashioned, as we have suggested, from hybridity and diaspora but which calls up authenticity and in order to do so returns to the indigenous peoples. However, the indigenous peoples of the 1990s are a politicized collective subject no longer willing to be subaltern voices and inserted into agendas that they do not make.

In another way, too, the issues of time/space have been modernized in the development of Ecuador as a nation-state. In an interview with Luis Macas, the leader of CONAIE, he wryly commented upon the early explorers in search of the centre of the earth: they located Mitad del Mundo (Quito) in place whereas the indigenous peoples located it in time. This is in part a consequence of the not uncommon articulation of space/time in many languages including Quechua and expressed in the word/concept *'pacha'* (Skar 1981: 48). Importantly, Freddy Ehlers' re-visioned history of Ecuador did not include the African diaspora. The notion of racial time in relation to the African diaspora was signified and is signified by an absence, a collective amnesia in relation to the history of people of African descent in Ecuador. The notion of racial time is an important one suggested by Winant (1994) in relation to the development of capitalism and the world economic order but which has major repurcussions in the post-modern world. Modernity was fashioned within a racialized time which has been submerged within an account of capitalist development which has concentrated upon economic relations and has not been understood as racialized. However, the entire history of modernity and the development of capitalism are bound to the era of enslavement which ended in Brazil only in the 1880s. The African descent populations have a different account of time via their history, a time so powerfully explored in Toni Morrison's novel *Beloved* and evoked by the African-Ecuadorean writer, Nelson Estupiñán Bass in *Curfew* (1992).

Popular culture articulates space/time in relation to the nation through the calender of popular festivals and their location alongside the national events which, although enacted in the capital city, Quito, often have local variants. In this way the local and the national are brought together. Equally, towns and cities celebrate their own festivals similar to the Quito festival; these were often cited in the interview data as the main ways in which individuals participated in local life. For those living in the poorest *barrios* their identities were constituted in relation to the naming of the *barrio* for the day of the original land invasion as in Guayaquil and Esmeraldas. Named for the day of the invasion, the *barrio* locates itself firmly in the present and in place. For example, *Barrio veinte ocho de Mayo*, a working-class *barrio* in Esmeraldas settled in the mid-1970s is now a popular residential area to which poorer people aspire. The day is celebrated by local people to mark the founding of the community and participation signals community membership.

National time and conceptions of space have equally been reframed within the alternative national projects of the indigenous movements, with its popular nationalisms that draw upon difference and the multiple identities of indigenous peoples. It is a complex task to generate a politics which provides fusion and difference simultaneously and the issues of time and space are crucial to this venture. The CONAIE organization in Ecuador has shown itself to be one of the most successful of the indigenous organizations in generating from diversity a collective subject that can intervene in powerful ways in the political life of Ecuador. In part this is a product of the re-framing of time and space which calls up a specific subaltern history in relation to the importance of place and land (as explored in Chapter 5). But space also plays a pivotal part in the de-nationalizing of an indigenous identity, calling up a history that transcends modern national boundaries; spatially the horizon is transnational. The internationalism of the indigenous movment is, however, place-centred in ways that the internationalism of the Left did not exhibit, located as it was, and for some still is, with a global class identity.

As Castañeda's (1994) analysis of the Latin American Left emphasizes, the Left has always been mindful of the unfinished business of the nation, and has sought ways in which to ally the project of socialism to a populism dedicated to the reintegration of the dispossessed through a new national project. While this has privileged a discourse based around class analysis, ethnicities and racism have not been foregrounded. Instead, there has been a powerful appeal to the 'other' as an externality defined in practice by the USA and, not surprisingly, characterized as malevolent in intent and action. The USA formed the exterior against which the interior of the nation could be framed but the Left has also had to place the increasingly well-organized and articulate indigenous and environmentalist groups. At one level it has not been difficult to reframe these groups and their politics within a discourse on the nation which foregrounds 'our people' and 'our land/resources', part of a reconstituted history and society in which ethnicities have a major role but this has been little acknowledged by the Left.

Clearly, this is part of an ongoing story of 'American' relations in which forms of popular culture and the mass media have an increasing impact in framing a vision of the contemporary world. While politicians and capital across the Americas acknowledge and seek to profit from the growing power of media forms and consumerist accounts of popular culture, both socialists and the indigenous movements have sought to distance themselves from these developments. There are historical reasons for this: an anti-US imperialist stance, the affirmation of indigenous traditions, cultural specificities and the revolutionary tradition in Latin America. Consequently, as Green (1991: 94) suggests, 'The mass media has also become a battleground in the struggle to define Latin America's identity'.

The impact of US films, television, radio and youth styles is evidenced throughout the cities of Latin America but should not be read simply as the latest in cultural imperialism. The story is a much more complex one. Thus, the writer Gabriel Garcia Marquez has decided that he should engage with popular culture and his next project is going to be the writing of a *telenovela*. *Telenovelas* (soap operas) and football are the most potent elements of mass popular culture and each form, in its own way, both reproduce and generate the social imaginary of the nation. This relates, in part, to the foregoing discussion of time and space because both football and *telenovelas* provide a fusion of national time/space, explored below. It also speaks for the power of mass media forms in the everyday lives of people in the Latin American states. For example, it is impossible to ignore the cultural impact of Xuxa and Globo television in Brazil, now the fourth largest television company in the world; more people in Brazil have televisions than have running water and refrigerators. It is also the case that both Globo television and the Mexican company *Televisa* export television programmes as cultural commodities, especially *telenovelas*, across Latin America and to Spain and Portugal. The enormous impact of television in Latin America has not, however, created a passive audience (Mattelart 1979). Instead, active engaged viewers, drawing on a world which they know and value, generate what Martín-Barbero (1993) has called 'mediations'. Mediations are the outcomes of the encounter between mass media forms and the cultures within which receptions take place often generating hybrid, cross-over forms. *Chicha* music in Peru is one example which combines Andean musical forms with electric guitars within the urban milieu of the poorer parts of Lima. Meanwhile in Colombia the singer Carlos Vives combines rock styles with the rhythms of the Pacific coast to produce music popular throughout Latin America, the USA and Europe.

TELEVISUAL NATIONS

Telenovelas, watched by millions on a regular basis, have been the subject of considerable discussion, especially in relation to the gendering of popular culture and the radical/populist potential that is suggested by a genre which is narrative, easily read and polysemic (Martín-Barbero and Muñoz 1992). In Ecuador *telenovelas*, imported from Mexico, were prime-time television, rivalled only by football for audience ratings and popularity; in May 1994 they ranked first and second. Based on romance, the *telenovela* is powerfully melodramatic, building the stories of nations and states into the narratives of families complicated by numerous subplots. *Telenovelas* are, therefore, complex cultural productions, a carnivalesque form, suggests da Matta (1985: 96), where 'the author, reader and characters constantly change places'. With major corporate investments at stake, the concern for the economics of popular cultural productions has been well documented, coupled with an overriding concern with cultural imperialism, for example *How to Read Donald Duck*, (Dorfman and Mattelart 1975). As Rowe and Schelling comment (1991: 107) there was an assumption in much of this analysis that the viewers were passive recipients of US cultural imperialism which is now contested (Martín-Barbero 1993). In a more recent analysis the language of the debate has shifted from the language of imperialism to that of orientalism following the work of Said (1978), and the suggestion that in analysing mass media forms and cultural products it is important to be alive to the possibilities of 'techno-orientalism' in relation to the export of Euro/US media products as specific discourses (Morley and Robins 1995). This, however, provides no acknowledgement of the power of Brazilian and Mexican television companies and the role of national television networks in countering processes of 'techno-orientalism'.

Rowe and Schelling, like Martín-Barbero, trace the roots of *telenovelas* to earlier folk-based stories and songs which continue and also have their expression in the romantic stories published as *fotonovela* (comic book with photos) and *historieta* (comic-form). The grand themes of the telenovela bring together families with the histories of nations and states and some, like the recent *Café* (1994) in Colombia, make this very explicit.

Café used all the classic conventions of *telenovela* – a wealthy family and the daughter of an alliance between a domestic and a son of the house. Their lives were intertwined, generating a complex interplay of the social and the personal which made for riveting television watched by millions of viewers. Desire and sexuality were key ingredients as the plot in twists and turns relived the past through the characters, including an English man in the coffee trade. The issue of sexuality was foregrounded by the plot, which hinged on the ability of the present generation of sons to produce a male heir in order to inherit the family wealth. It was proving a difficult task and the *telenovela* presented a series of discourses on male sexuality which were organized around the main character and his inability to produce an heir. Thus, a familiar theme was also moving the discussion of sexualities across boundaries and away from women towards men. The dilemmas remained unresolved, a necessary pre-condition for the success of the programme. However, the issues of sexuality, class relations, imperialism and the nation sustained the drama.

It was also interesting that at the same time (1994) in Colombia, *Ana Luisa*, a low budget *telenovela* was produced on an alternative channel which had as its central character a woman, Ana Luisa, who left her husband and children to set up home with a younger male lover. Ana Luisa was shown to be struggling to maintain her relations with her children and neighbours against the condemnation of her family and, most especially, her female relatives, friends and neighbours. *Ana Luisa* presented the micro-politics of women's lives and the relations between the personal and the political in graphic ways. *Ana Luisa* was much discussed by women despite its lack of gloss and sophisticated production techniques. *Café* and *Ana Luisa* provide examples of the ways in which the progressive elements of this popular form enter into the public realm and shift the debate. Clearly, the substance of these productions is highly gendered with masculinities and femininities given maximum exploration within story lines that keep the viewers hanging on to the next episode.

In Ecuador during the early 1990s one of the most popular *telenovelas* was *Guadalupe*, a Mexican import, that again used all the conventions of the genre. The central character Guadalupe was the daughter of a beautiful but maltreated domestic within a wealthy family who sets out to avenge her mother and restore her honour, first, via marriage into wealth which proves disastrous, and second, via success in business which (despite any depiction of what might be construed as a working day) seems to succeed. Guadalupe travels from crisis to crisis representing, the narrative suggests, for all women the contradiction between independence and emotions; she comes through these crises as a single parent secure in her self. The presentation of a divorced woman claiming her own space and autonomy provided an important corrective to moralizing discourses on motherhood and marriage thereby challenging an account which has already been, in part, superseded by the lived experience of many women's lives. In addition, *Guadalupe* tackled rape and physical abuse by men; dramatic narrative highlighted the position of women and the power of men. The terrain for this discussion was a wealthy, upper-middle-class milieu, an important setting belying the fiction that violence against women is an issue solely among the poor or working class. *Guadalupe* also offered the position of wise women to the servants of the household who provided a folk wisdom and ongoing commentary on the decadence and foolishness of the wealthy. As popular television it was enormously successful as it was visually exciting and packed with passion and drama that related to issues constructed as fundamental rather than trivial. The audience was not patronized but invited to be part of a debate on serious contemporary issues, because of the lack of fixity between viewers, readers and characters, constructing a space of dissent into which people could insert themselves and contribute as knowing-subjects. This 'space of dissent' also organized an articulation between the home as the site of consumption and the national. The two became one through the correlative imaginaries of viewer/subjects discursively constructed via the ensuing debates. The importance of this in relation to nations and national identities cannot be underestimated because these processes test out specific commonsense understandings in circulation, like the issues around male violence generated from *Guadalupe*, and rupture the 'naturalization' of accepted views. In so doing people see themselves and their views in the mirror of the nation.

Guadalupe in Ecuador, as with *telenovelas* throughout Latin America, provide a collective space for the nation in several ways linked to the discussion of national space/time. *Telenovelas* like the Colombian *Café* and others from Brazil, Mexico and Argentina offer a vision of the historical and social development of a specific nation-state drawing the viewer into a story which becomes a form of cultural capital and which interacts with historical time and the geographic and social space of the nation. This was transparent in *Café* where viewers could immediately connect with the sign *Café* as a signifier of Colombia and the political and economic importance of coffee in the past, present and future of the country. Yet these dramatized histories were organized around production values which emphasize visual impact and human interest stories, basically seduction and betrayal in love or business. Nevertheless, *Café* offered a 'telespace' shared by millions at the same time week after week, in which Colombians engaged with a vision of their past and present society – a powerful form of interpellation. Despite little attention paid to the plantation workers, issues of exploitation and racism, it is possible to view the main narrative as a tale in which the poor and dispossessed (embodied in the main female character Carolina) were 'represented' and shown to be active in their attempts to right injustice and claim a place in the nation. *Telenovelas* offer a narrative and characters that are shared by millions and that are referenced in other media forms. This form of intertextuality further reinforces the sense of a national story and identity, a sort of living history of the nation in which everyone, via television, can take part. Ecuador is an interesting case in relation to these observations, because it has not had a national *telenovela* through which to tell the national story. Instead, *telenovelas* are imported, currently from Mexico, and previously from the southern cone states. This does not mean that the ways in which *telenovelas* provide a space for the national are denuded but that the millions who watched *Guadalupe* engaged with a Mexican *telenovela* made largely on location in Miami. The ways in which *telenovelas* bind so many people/viewers in time and space via television is an example of how national time becomes generalized across the nation. This is further reinforced by the inter-textuality which engages viewers as readers and provides a world within a world shared across regional and class boundaries.

But the interpretation and reinterpretation of this world fractures by class, ethnicity, region and gender. There has been considerable discussion of the engendered and particularly feminine quality of soaps and *telenovelas*. While they certainly present a world in which the power of women and the representation of the feminine are major motifs, *telenovelas* are not viewed and enjoyed by an entirely female audience. It would be too simple to assume that women watched *telenovelas* and men watched football. Certainly, more men than women watch football in Ecuador, but *telenovelas*, while seen and analysed as a women's genre, have audience figures showing their general appeal. However, the issues, some of which we have alluded to, generated and sustained in *telenovelas* did speak to issues close to women, the configuration of femininity and sexual politics. Further evidence for the generality of interest in *telenovelas* comes from our own research into the television programmes watched by men and women in different regions and classes, and the market research data for Ecuador. Middle-class men in the major cities when asked in the interviews we conducted were keen to persuade us that their television viewing was concentrated on news, documentaries and sport. But this was clearly contradicted by the market research on viewing figures in May 1994, which gave the highest audience figures for these groups to both sport and *telenovelas*. This is an important corrective to the widely held view which we in part shared at the start of our research that football was a powerful national but masculine story whereas *telenovelas* were a powerful feminine one. The data also underline the importance of mass media forms, most especially television, in the lives of peoples and nations. Television in this way performs the task of horizontal integrations within the nation-state via the news and current affairs programmes and also via popular television exemplified in *telenovelas* and football. But this integrative work is a constant process of 'mediations', tentative, unstable and lacking fixity.

The ways in which television through documentaries, news coverage and these popular forms is able to provide a national time/space and constitute national and cultural identities is clearly articulated by Donald (1988):

> the apparatuses of discourse, technologies and institutions (print capitalism, education, mass media and so forth) which produced what is generally recognized as 'the national culture' . . . the nation is an effect of these cultural technologies, not their origin.
>
> (Donald 1988: 32)

Moreover,

> This cultural and social heterogeneity is given a certain fixity by the articulating principle of 'the nation'. The 'national' defines the culture's unity by differentiating it from other cultures, by marking its boundaries; a fictional unity, of course, because the 'us' on the inside is itself differentiated.
>
> (Donald 1988: 32)

It is also undermined by the globalization of cultural products which suggest different trans-national constituencies.

Private broadcasting has always to look to sponsors and advertising revenue but in order to secure this it has to generate high audience figures. In Ecuador there was considerable competition between a number of channels. One channel was an 'educational' channel which developed debate on national questions and priorities, most especially those concerned with the indigenous movements and Amazonia. In one programme, the government and managers of the major oil companies were interrogated and called to account by the local population in the areas where oil was drilled and transported. Television also had a regional flavour, especially between the Costa and Quito reproducing the political and economic rivalries between the two areas. Diversity is an important part of the mediations which occur in relation to cultural products and from which a national culture is forged, as Donald (1988) suggests.

Regional differentiation can also been seen in the use of radio which offers both a national account via world news and events within specific nation-states, and a cheap and universal medium for political, cultural and religious groups. Radio is much in evidence in Amazonia as a core medium of communication which the recently arrived Protestant and Messianic religious groups have expanded through new radio stations. The power of radio to generate and sustain communities in struggle and produce a sense of belonging is exemplified by Radio Venceremos in El Salvador (López Vigil 1995), which sustained the guerrilla struggle through the decade of the civil war, and also in Mexico, where the Zapatistas captured the local radio in January 1994 and turned it into 'Radio Zapata'. The station's first broadcast was a declaration of war against the federal army (Katzenberger 1995). Radio as a cheap and available technology of communication allows for diversity and provides a powerful counter-hegemonic tool bringing together politics and popular cultures in often novel and innovative ways.

The diversity of Ecuador can also be seen in the ways in which it is very difficult to produce an agreed literary or artistic 'canon'. A great variety of writers, who clearly speak differently to different sections of the population, were represented in the interview data. No one single figure, like Gabriel Garcia Marquez in neighbouring Colombia, exists, and while the school syllabus provides definitions of *patria* and *nación* there is no attempt to define a literary nation. However, names like Juan Leon Mera (who wrote the national anthem) and more currently Nelson Estupiñán Bass (an African-Ecuadorean from Esmeraldas) do reoccur in popular knowledge of literature. It is, like so many other facets of Ecuadorean nationhood, a plurality of possibilities. It is, however, male authors who are more within the national imagination.

The mass media forms and especially television occupy this space of the social imaginary in very powerful ways, offering Ecuadoreans a vision of themselves and of Ecuadorean identity which is both consolidated and contested within television programmes and re-visioned by the diversity of the viewers. The power of mass popular culture has not been so easily understood by either the Left generally or the indigenous movement in Ecuador as it has been by the Right and the current generation of politicians. The cultural moment for these groups has been set up in opposition to mass popular culture. For example, CONAIE organized an indigenous, first nations, video and film festival using some powerful historical and contemporary material from the archives to the current moment of the 1990 uprisings. Sadly, few people attended the event in Quito; the reason for the poor turn-out was acknowledged to be a clash with two important televised football matches that were part of the World Cup qualifying rounds.

In an earlier clash with football, in the summer of 1993, CONAIE leaders, frustrated by their discussions with the government on land reform, walked out of the discussions and threatened to disrupt the Copa América (Latin American soccer competition held every two years) which was being hosted by Ecuador. This pronouncement was greeted with derision by a hostile press as an example of the distance between the indigenous movement, the popular concerns of Ecuadoreans and national pride as the hosts of the Copa América. The coverage allowed the media to negatively construct difference in relation to the indigenous movement and to distance the leadership of the movement from the 'people'. CONAIE, however, is too powerful and confident to be troubled by negative press coverage but it did encourage some of the leaders of the movement, themselves football fans, to reconsider the role of mass popular culture in the struggle for indigenous rights.

FOOTBALLING NATIONS

Football is a passion throughout Latin America and no less so in Ecuador. Low-wage workers in Guayaquil, home to the successful Barcelona team, are willing to miss the main meal of the day in order to be able to afford a ticket for the match on Saturday. Football matches are alive with music, banners, songs and chants, giving the event a festival atmosphere full of laughter and energy. It is this joy in football that was imported into the USA for the 1994 World Cup with Mexican waves, carnival-style Brazilian costume, songs and drums adding to the street party. Football is also megabucks, politics and corruption so tragically played out in relation to the Colombian national team, one of whom, Andrés Escobar, was shot following a disastrous performance in the 1994 World Cup. Life for footballers is gentler in Ecuador but power and money in relation to national pride is a highly charged cocktail.

Ecuador is not one of the major footballing nations of Latin America; it has yet to produce a team with the flair of the ill-fated Colombian team or individual players of the brilliance of Pelé or Maradona. However, one of the teams from Guayaquil, Barcelona, has done well in Latin American tournaments and there is no lack of enthusiasm and passion for the sport. When Ecuador was knocked out of the World Cup qualifying rounds at an early stage there was deep mourning and extended media coverage of the defeat and the reasons for Ecuador's performance. National pride was at stake and large sections of the population mourned the end of Ecuador's World Cup career. But Ecuadoreans cheerfully and generously shifted their loyalty to Colombia and pinned their hopes on their neighbour, especially after the Colombian team beat Argentina 5–0 in Buenos Aires.

Football loyalties show clearly the factions and characterizations that feed a sense of national identity. The Argentines are considered arrogant, rude and racist especially after they called the Colombian black players and the team in general a bunch of 'negritas' following their defeat. The language here is gendered suggesting a girlish, feminized quality of the Colombian players, adding to the racist insult. Such insults fuelled Ecuadorean support for Colombia. Yet in the play-offs the Ecuadorean sports commentary used insults to distance Ecuador from other nations, calling the Uruguayans 'animals' (ibestia!) in relation to their tactics, followed by an analysis of their unsporting behaviour compared to the prowess of rule-bound Ecuadoreans. Watching this particular game left some doubts on this but, again, national pride was at stake and the commentators clearly had the mood of the viewers. Those with whom we watched described the Uruguayans in the same way and maintained a running commentary on their manipulation of the rules. These matches (and those for the earlier Copa América in 1993) were, according to the audience ratings, watched by 90 per cent of Ecuadoreans; the matches were followed by front page banner headlines in the press. The Copa América was important because Ecuador was the host nation and nothing could detract from this even if many of the matches did not go in Ecuador's favour. It was an honour to host the matches and demonstrated the serious commitment to soccer by the government and peoples of the country, putting Ecuador in the centre stage of Latin American football for the duration of La Copa.

La Copa generated a series of exciting matches and a surfeit of newspaper and television coverage on the game, the managers and the star players but also, of course, on inter-nation rivalries and the place of Ecuador in Latin America. Football is the 'national' game in Ecuador and the team that takes the name *Nacional* is, interestingly, a military team set up in the 1970s by the governing military regime. The most successful national and international team is Barcelona, one of two teams in the coastal city of Guayaquil. The original Guayaquil team is Emelec, wearing blue and white and supported by those who feel that they are the true Guayaquilaneans and the city fathers, whereas Barcelona (red and white strip) is more populist, reflecting new money with a huge stadium used by one politician for rallies. One-time president and long-term politician, Leon Febres Cordero, was quite clear that football and national politics are bound together. He recognized the symbolic power of being associated with the popular club both in terms of support from members and fans, and the success and prowess of the team. In Quito, too, noted politicians are attached to the major football clubs and their fortunes are often as unpredictable.

Clearly, local and regional rivalries are played out through the major teams and it is a source of satisfaction to the Costa that their team is the star team in Ecuador. Football not only reinforces class and regional loyalties but also produces them, offering symbols of the cities and localities. In a larger way football offers a series of cultural referents shared by many, especially men in Ecuador, and elsewhere, who play the game in an amateur way as well. Importantly, the star players are often Ecuadoreans of African descent. During 1993 and 1994 half the *Nacional* team was of African descent offering success, status and recognition within the national realm to Ecuadoreans of African descent, which they struggle to enjoy elsewhere in the country. It is, of course, a familiar story of black footballers, well known in Latin America, Europe and now the USA. Black footballers suffer racism on and off the pitch and no less so in Ecuador than other countries around the world.

Football offers a clear focal point for national discourses as football teams have come to symbolize national pride and shame. These are deeply felt emotions tied to the conceptions that individuals have of the life of the nation and some hazy notion of 'national destiny'. What this means for nations and national identities in the modern era is that football, too, contributes and marks conceptions of national time and space. The game of football is itself a product of modernity bound to disciplinary modes that have been codified in the twentieth century. The discursive space of the game is rule-bound via local, national and international institutions. It is the global game played across the world having been transported initially from Europe and re-visioned as it has travelled around the globe.

In the Latin American context it was, suggests Mason (1995), British sailors who initially played a game something like modern football in the ports of Latin America. This beginning was developed through the British communities, especially in Argentina, where football – allied to the development of schools and leisure – was played from the end of the nineteenth century. This was followed by touring teams like Exeter, who spent a month in Argentina in 1914. By 1917 there was a South American championship and the development of the game was assured by the enthusiastic response of the crowds. South American teams were developing very quickly and in the Olympics of 1924 Uruguay beat most of the European teams to carry off the World Cup. It was a great moment and a national holiday was declared in Uruguay. Seventy years later in 1995, Uruguay won the Copa América held in Montevideo. Like Argentina and Brazil, Uruguay has continued to be a top footballing nation with teams, year after year, that have made a huge impact on the world of soccer. Part of this impact has related to the ways in which, having been introduced to the game by the British, these nations then re-invented the game in stunning ways that were imported back to Europe.

As Mason's account of the development of football in Latin America details, the rise of the professional footballer was, in part, a response to the growing numbers of Brazilian, Argentine and Uruguayan players who were being recruited by European teams. Fluminense in Brazil started paying players in 1932 but their great rivals Flamengo opposed this move initially. Mason (1995: 51) quotes the president of Flamengo: 'a gigolo who exploits a prostitute. The club gives him all the material necessary to play football and enjoy himself with the game and he wants to earn money as well? Professionalism degrades the man.' The emotive language and the issue of professionalizing the game were a source of contention throughout the soccer world but the trajectory of professionalization could not be stopped by appeals to the idea of the gentleman's game. Football was now an international affair within the growing power of capitalist development that sought ways in which to commodify sport through the developing leisure industry and mass media forms. By the time Argentina won the South American Championship in Ecuador in 1947 football had become 'the passion of the people' throughout the states of Latin America and in many parts of the world.

In an interesting analysis that uses football as a major purveyor of national sentiments, as we do, Archetti (1994a) traces the ways in which an 'arena of freedom and creativity' was developed within the Argentine game which separated it from the parent game imported from the British. Instead, from the 1920s a much more balletic style of football was developed which Argentina could call its own, and which came to mark Latin American football more generally. In the Argentine case Archetti sees the invention of the style as akin to Hobsbawm's (1983) 'invention of tradition', a cultural reconversion which offered symbols, myths and a distinctive 'national male style':

> Football . . . produces an overlap of practical and symbolic constructions of national characteristics, national values and national pride and sorrow. The 'national' will thus come to be perceived as 'naturally' masculine. Women excluded from active participation in this site of national construction can, however, 'identify' themselves with the 'team as nation'.
>
> (Archetti 1994: 236)

There is plenty of evidence that women do indeed do this while recognizing their exclusion most especially, and interestingly, from the emotional space of the game.

The discursive space of football is a site in which men can express their emotions while emphasizing the public face of masculinities. This emotionality is also often expressed in the language with which footballers and fans discuss the game. One pertinent example of this, which relates to Argentina and the development of a national style, is provided by an account of Real Madrid (Melcon and Smith 1961). It includes an interview with the great Argentine player Alfredo di Stefano called 'For the Love of Football' in which he celebrates the English team:

> I do not support the view that English dominance of world football is at an end. . . . England has all the elements required in a great footballing nation. . . . South American players are usually gifted ball players, . . . it is almost a gift of their environment, but they are careless of physical condition, . . . and they like to romance with the game, rather than approach it with a thoroughly professional determination.
>
> (Melcon and Smith 1961: 48–9)

This view, of course, reinforces the notion of the stoical, professional white English player against the Latin of temperament and flair, exemplified following a friendly international match between England and Colombia at Wembley in which it was said of the gymnastically adept goalkeeper, Higuita, 'His game, like his life, needs exotic gestures and close shaves' (*Guardian* 7 September 1995). Reviewing the game, David Lacey opened with the comment 'England's redesigned team . . . refused to be hypnotised by Colombia's mesmeric football' (*Guardian* 7 September 1995). Such comments have been consistently recycled in British football, contributing to the Otherization of players who are not white and English (Westwood 1991). Di Stefano concludes his interview with the words: 'Football is a way of life having a special dignity. . . . There is no finer game than this one, which these youngsters, in their turn, will hold in trust'(Melcon and Smith 1961: 51). The passion with which football is described offers a glimpse of the level of emotional commitment expressed simply and elegantly by Pelé: 'Football, the beautiful game'. It is part of the romance of football that it can generate forms of poetry and eulogies which are comparable to the sentiments expressed towards the nation.

The popularity of football and its embeddedness within popular culture was not lost on the politicians and nation-builders of the post-Second World War era, especially in Argentina, where its symbolic power was used in the centenary celebrations as early as 1910 and 1916. As Mason (1995) notes in relation to the Vargas era in Brazil:

> The Vargas era was a turning point in the relationship between football and politics. From this time not only the Federal Government but individual politicians would try to associate themselves with what was becoming an increasingly powerful manifestation of Brazilian popular culture.
>
> (Mason 1995: 63)

Equally, in Argentina it was Juan Perón, assisted by Eva Perón, who systematized the relationship between state, nation and football to the point where in 1953 Perón declared 14 May to be 'Footballers' Day', following the Argentine defeat of England in Buenos Aires, a celebration to take place annually. Football has been used by governments as a distraction; generals like Pinochet in Chile supported football teams that are hugely popular, and 'in the national interest' have stopped players from going abroad. Mason (1995: 76) concludes: 'It is repression or the fear of it which keeps people politically passive in dictatorships, but football, like the television soap opera, plays a part'.

In a television profile of Diego Maradona, despite his obvious quarrels with the Argentine footballing authorities, Maradona said, 'I think it's everyone's duty to be a patriot' (*True Stories*, Channel 4, 19 May 1995). In relation to his own relationship with the national team, he spoke movingly of his pride in playing for the Argentine side in its triumphs and of his sorrow in defeat. The national story that is part of these emotions has been specifically promoted via football in all the major footballing nations in Latin America – Brazil, Argentina and Uruguay. Soccer in Brazil promoted national unity, seen again in the World Cup of 1994 (Lever 1986). The Brazilian success also promoted Latin American football against Europe and provided a source of pride across the continent. Watching the World Cup final in Ecuador, the normally quiet Quito exploded into a street party in which the chants of *Olé, Olé, Olé*, were ringing throughout the night. What was also interesting was the way in which the Brazilian colours and flags were soon replaced by Ecuadorean ones: cars hooted and the crowds shouted to the point where anyone not familiar with the actual result might just have thought that Ecuador had triumphed. The power of these identifications is expressed by Roberto da Matta: 'Football reflected the nationality, it mirrors the nation. Without football we Brazilians do not exist – just as one could not conceive of Spain without the bullfight' (da Matta 1990, quoted in Mason 1995). Da Matta has a very clear message for the doubters in his emphasis upon the power of popular culture, carnival, football and popular religion to provide sources of identity generated and sustained from the popular, by the people.

> After all it is better to be champion in samba, carnival and football than in wars and sales of rockets. . . . It shows we can love Brazil with its

hymn and its flag, maintaining our lucidity relative to the regime we want to transform.

(da Matta 1990: 132)

Mason (1995) is sceptical, in part, as his earlier comment suggested, because football leaves economic relations, corruptions and abuses in place. While acknowledging this, we would suggest that Mason does not fully comprehend the power and pull of the imaginary in a national identity symbolized by the national football team, most importantly where these teams have been as successful as Brazil and Argentina, and how crucially implicated football is in the modern project in Latin America to a degree not found in Europe.

This returns us to issues discussed earlier in the chapter. Football generates a sense of national time and space from the popular rather than the official histories of the nation. However, football can clearly bring the two together in the most manipulative way like the staging of the World Cup in Argentina in 1978, where the generals decided that football could be used as a unifying project in a fascist state. It failed, despite the wall built to hide the slums between the airport and downtown in Buenos Aires, and the attempt to hide the brutality of the regime. No one was fooled, least of all Argentinians who invented counter-slogans and a bumper sticker that depicted a football covered in barbed wire. It was bad news for the generals because it was at this time, through the media coverage, that the images of the Mothers of the *Plaza de Mayo* (a human rights group of women searching for disappeared relatives) were beamed around the world. As Kuper (1994: 178) concludes, 'The World Cup was bad for investment and tourism in Argentina but good for human rights'.

The national boundaries generated by football and the ways in which they are transgressed offer a symbolic space into which people can insert themselves and where their identities can be called up not only as individuals, but also as part of the collective subject of the nation. Located not just in an ethnographic present, this relates to a past of tragedy and triumph, games won and lost, trophies gained and national pride foregrounded. It is precisely because this is an emotional space, a romance, an affair of the heart that it is so powerful (Westwood 1991). The history of the nation is reinterpreted and dated by a series of World Cup successes in Brazil and Argentina. Thus, it is precisely in the arena of global, international competition that the game of football is so prominent and so clearly reinforces the imaginary of the nation. However, the game itself is internationalized and has been for some time, especially in relation to the trade in players across continents which brought extraordinary players like di Stefano and Maradona to Europe. Equally, while presented as a democratic and popular arena which belongs to all, the politics and economics of football show very clearly that the story is much more complex and that football is not simply nostalgia or 'just a game'. Football is deeply implicated in politics, capitalist commodification and profit maximization. Yet, it is still privileged as narrator of the nation: the narrator in question is masculine and football is a romance built around an idealized masculinity, as Archetti (1994a) elaborates in relation to Argentina. Even this, however, is full of contradictions exemplified in the life and times of Maradona, so different from those of Pelé. The social and ethnic backgrounds of these two men fuel the romance of football while simultaneously silencing issues of racism and class exploitation, generating national stories that produce national heroes from below. At the popular level, they mirror the processes of nation-formation with its horizontal integration and symbolic ownership of heroes. It is these processes that are in play in what we have called 'correlative imaginaries' where identities are placed in a space or site such as football on the basis of a correlation of the self with the social which imagines mutual intelligibility, even though the interests of managers, players, financiers and spectators may be economically and politically antagonistic. However, the processes of identification are not impeded because football is constructed as popular culture which is understood to be a realm of creativity and freedom. This is, in part, a fiction but a necessary fiction in relation to the ability of football to symbolize the nation and national destiny. Correlative imaginations place national and individual destinies side by side, the one within the other and, apart from war (that other major masculinist story of the nation), football is the most powerful evocation of a 'correlative destiny'.

RE-FRAMING POPULAR RELIGION

As Kuper (1994: 1) notes, 'More people go to prayer than to football matches'. The role of popular religion and its relations to other sites of popular culture, with which this chapter concludes, was illustrated by the visit of the Colombian football team to the church at Buga before the team left for their ill-fated World Cup games in the USA. Buga is a small town in Colombia now dominated by the church and its followers, who come in their thousands for healing and protection to a church famous for the relics of a miraculous cross. Such 'miracles' are celebrated further afield than Colombia, and Buga is allied with a popular Catholicism expressed in the services held in the church which include popular music, electronic messages and fighting fists held aloft in the air. The message is collectivist and international, politically populist in a revivalist way. Attending mass at Buga is an empowering and life-enhancing experience and, in this context, it is easy to embrace and to see why popular Catholicism, like popular religion more generally, is so seductive. Although post-modern writers from Lyotard (1984) onwards have emphasized 'the death of the grand narratives', the 'ones that got away' appear increasingly to be religious grand narratives which grow and flourish. However, they do not do so alone but in relation to a complex politics of ethnicities in the post-colonial era which, consistent with the post-modern turn, fractures the narratives, thereby generating multiplicity (see van der Veer 1994 on the rise of Hindu nationalism in India, for example).

That Latin America is Catholic is part of the general commonsense understanding of the formation and development of the region's nation-states from the period of colonization to the present day, but it has never been uniform and homogeneous. From its arrival in Latin America, Catholicism brought with it the internal divisions of the faith and its organizational forms. In addition the doctrine and practices were 'indigenized', creating a hybrid Catholicism generated from extant beliefs and practices that constantly engaged with official accounts of the religion. Given this background and the politics of Latin America it is not surprising that Liberation Theology, bringing together certain strands of Marxism and Catholicism, should have developed in Latin America.

The precursor to Liberation Theology may be found in the migration of Catholic missionaries to Latin America in the 1950s and 1960s, some of whom became major exponents of liberationist views, and in the forerunner, social Catholicism, which sought to incorporate the poor and the working class in a consensus capitalism on a model of distributive, but unequal, justice. Liberation Theology went further, incorporating a new vision for the gospel, a 'commitment to the liberation of millions of the oppressed of our world' (Boff and Boff 1987: 8f). More radical views have come from worker-priests and priests who have given a clear commitment to some of the political liberation struggles in the region, in Brazil and Nicaragua, for example. The danger for theology in these interventions is that the liberationist message becomes secularized and located with specific political and social movements divorced from Catholicism. Liberation Theology was, in part, a response to the structural developments and inequities of globalized capitalism as part of modernity, and it continues to wrestle with the contradictions within these developments.

In many of the states, and reproduced in Ecuador, there is an uneasy truce between the two main wings of the Catholic Church, Opus Dei and the Liberationists. Both now use the language of the 'Option for the Poor' following the impact of the Medellin declaration on 'Justice' and 'Peace' in 1968 which encouraged a Catholicism that was engaged with the material struggles of the poor. This 'new' Catholicism inadvertently entered the political terrain through developments linked to the Comunidades Eclesiasticas de Base (CEBs: Christian Base Communities) which, most especially, brought women together in pursuit of educational and social goals. The Christian Base Communities were a vital part of the sustenance of civil society during the most repressive years of the 1970s and 1980s in many countries. Among other issues, the CEBs provided a platform for women in relation to political goals that are still nurtured within these groups (Jaquette 1989; Machado 1993; Corcoran-Nantes 1993). When we discussed this with the foremost exponent of Liberation Theology in Ecuador, Monsignor Tobar, he insisted, 'women are the Church. There would be no Church without women'. He spoke poetically of the struggles of the poor for land, education and freedom; he could not envisage a Church standing aside from these struggles. Monsignor Tobar is an important figure and one moment in the complex configuration that is the Roman Catholic church in Ecuador. He co-exists with an articulate and well-organized Opus Dei fraction who occupy positions of power and have a major role in the Universidad Católica (Catholic University) in Quito, the capital city of Ecuador, as they do in universities in Spain and the other Latin American states. Opus Dei speaks for the embeddedness of Roman Catholicism in the political institutions and cultural life of the nation-states of Latin America. But this solidity belies a sea-change in religious affiliation which marks Latin America from the southern cone to the Andes, with the profound shift towards Protestantism in the region (Stoll 1990). There are an increasing number of sects, evangelical Christian churches and Protestant churches that have been, and are, recruiting in ever larger numbers through-out the region.

Framed by a Gramscian understanding, religion can be analysed as both part of the generation and sustenance of hegemonic relations and as contested terrain in which counter-hegemonic understandings and an alternative commonsense can be forged. It is a contradictory story:

> In general, religion has been studied from critical perspectives as a significant part of 'the will for political domination' associated with established political power. However, there are now criteria, expectations and concrete processes in various Latin American countries that invite us to consider religion as a significant part of the 'will to transformation'.
>
> (Torres 1992: 105–6)

Torres views the current crises in the churches in Latin America as part of an ongoing crisis of the hegemonies within the different nation-states of the region, bound to the changing political and economic relations that have, from the 1970s, shifted the terrain in which the Church operates. The calls from Pope John Paul for unity are an expression of the recognition in Rome that the Catholic Church in Latin America is fractured. Pope John Paul has consistently reiterated a conservative view of the role of the Church and its teachings, at odds with Liberation Theologists and the oppositional stance of sections of the priesthood. This major fracture is reproduced in Argentina with an integralist Argentine Church allied with the armed forces from the 1930s into the 1970s and early 1980s during the worst excesses of state terrorism. But it was precisely in relation to these abuses that the oppositional elements within the Church were most vocal and well organized. Following the period of Perónism which reorganized the relationship between the Church and the popular classes, the new configuration created a split between the Church and its anti-Perónism and the masses of the people who identified with Perón and the practices of Perónism. Currently, 'the Church will continue to be an important traditional intellectual rooted in Argentine politics . . . in either the consolidation or – its opposite – the destabilization of Argentina's fragile democracy' (Torres 1992: 177).

The important role of the Catholic Church, with all its contradictions, in the generation and sustenance of national ideologies, is a pattern reproduced across the region, expressed in the notion that to be Latin American is to be Catholic. This has important implications for national identities constructed as imaginaries which include an adherence to Catholicism. Thus, people throughout the states of Latin America are interpellated as nationals in part through Catholicism as the national religion. These processes are not simply top-down inductions via schooling, public ceremony and official discourses of the nation but are part of everyday life in religious practices, rites of passage and celebrations within communities. Catholicism is in this sense embedded in national identities in countries like Ecuador and provides an important, but fractured, site for the production of correlative imaginaries in relation to the nation, in part due to the powerful emotional appeal of religion and religious sentiments. Equally, the inclusivity offered by the elision between state, nation and Church is reinforced through the powerful way in which the imagery of the religion forms part of the national heritage and memory, for example, miracles and visions like the Virgin of Guadalupe in Mexico. Such fracturing further reinforces the notion that religion, while representing a form of traditional intellectual in one site, is part of a de-centred hegemony which allows religion to be oppositional in another. The implications of this for national identities lie in the ways in which religion and ethnicities are interwoven. Thus, part of the ways in which ethnically distinct groups are 'Otherized' and excluded from the nation is in relation to their distance from the official version of Roman Catholicism, for example, the Jewish communities in Argentina or peoples of African descent practising diasporically constructed versions of African ritual and belief.

Disaffection from the Catholic faith is an old story, one which is represented in the devil stories so common throughout Latin America. Devil stories among migrants to Cuenca in Ecuador are used as moral tales and as a way of instilling appropriate values in children within the corrupting climate of the city (Miles 1994). The Devil represents the corrupting influences of capitalism and money which are hidden through the seductions that lure people into greed and selfishness. What is also interesting in relation to the stories is that the priest is a very common figure often shown as ineffectual, stupid and himself corrupt and unable to deal with extraordinary events or matters of life and death. Figurative priests have no knowledge and wisdom to impart, no compassion or strategy. By contrast the latter attributes are embodied in old women or local men who have a folk wisdom which usually saves the day; clearly juxtaposed is the popular 'really useful knowledge' in contrast to Catholic dogma and the power of prayer. The stories provide a critique of the piety of the priesthood and the hierarchical nature of the Church and in themselves express a disaffection with Catholicism which has a long history. This disaffection is reproduced among sections of the black population in Brazil and indigenous peoples from the Andes to Amazonia. Burdick (1992) suggests that Brazil accounts for half of the evangelical population in Latin America and this has important consequences for the ethnicized character of the movement to Protestantism. The appeal of Pentecostal churches in Africa and among the populations of the African diaspora is well known and it would be useful to explore further the ways in which these churches offer a cross-national reference among people of African descent in relation to an imaginary community of believers.

This issue of the relationship between nations, national identities and the hegemony of Roman Catholicism has taken a new turn because Catholicism, challenged from within and by the development of new political configurations, has also to deal with the success of the Protestant sects from without.

In the last redoubt of the Roman Catholic empire, where almost half the world's Catholics reside, we are witnessing the early beginnings of a new

Reformation. . . . Pentecostalists, Evangelicals and Mormons move from country to country and draw upon peasant, working class and lower middle class support. But, they are not content to place themselves at the margins. Instead, they are to be found entering the political fray, buying or building prestigious new edifices to God.

(Gott 1995: 6)

The politics of these groups is uncertain and highly differentiated but are largely seen to be aligned with conservative, traditional values. Those sects on the side of 'traditional values' could potentially insert themselves into the politics of the nation in ways that disrupt the monopoly currently exercised by the Roman Catholic Church in Latin American countries, turning believers into a political lobby in which the cultural contours of the nation could be foregrounded.

However, if the progressive potential of the Protestant sects is acknowledged, the impact appears to relate most importantly to gender politics. For example, it has been suggested that women are especially warm to the Protestant sects because they encourage family values in a more active, egalitarian way, emphasizing the responsibilities of men as fathers and husbands. Many of the new churches are also notoriously anti-alcohol, while more importantly they often emphasize the value of all believers and the equality of women and men in relation to faith, thereby undermining the commonsense of *machismo* (Brusco 1995).

The move towards the Protestant churches is generally aligned with the way in which Latin American states and peoples have embraced capitalism US-style, via the culture of consumption to be seen in the shopping malls, the baseball hats and Mickey Mouse icons, with major implications for conceptions of national and trans-national identities. In this context the Protestant message is consistent and offers both a way into the north and its modernity, and a means of staying there. The Chilean sociologist Claudio Véliz (1992) suggests that the evangelical churches are a form of a 'cargo cult' behaviour' 'that attributes the capacity to sustain a prosperous good life – as well as generate attractive consumer goods and cultural artefacts – directly to the culture of industrial capitalism' and it is this culture that is bound to 'the tenets of evangelical Protestantism' (quoted in Gott 1995: 27). Confirming this view, a study of conversion among Highland Guatemala Mayas emphasizes the shift in economic understanding among converts: 'Success in life is increasingly defined in economic terms and in judgements of good and bad investments interpreted broadly' (Goldin and Metz 1991: 336). These shifts in ideology and practice are part of a larger transformation and may affect non-converts equally. In this sense a new common sense is being fashioned, within which the Protestant churches organize an account of the social and economic changes underway in the Latin American states which offers a place to many who are on the periphery. The Protestant sects place adherents within the frame as participants rather than as onlookers, outside the processes of change.

An alternative language within which to place these processes suggests that the Protestant sects are part of the alternative modernities developed in

Latin America, 'indigenized' modernities in which the globalization of more specifically US capitalism has come to shape the whole region. The Protestant sects offer an individualized self which meets modernity, and its discontents, in relation to specific disciplinary modes internal to the sects. But these 'technologies of the self' also produce disobedient subjects of the state/Catholic Church/nation complex in forms of compliance and resistance simultaneously. Equally, like the US version, these sects offer a space of belonging in a world of transition, the parameters of which are clearly drawn, suggesting a sense of safety and intelligibility in a world where the older cognitive maps no longer seem able to offer a route.

Conversion to the Protestant sects among migrants from an Andean Peruvian village, Tapay, illustrates these themes (Paerregaard 1994). Migrants returning to the village have been central to the conversion of villagers, viewing this as part of an attempt to modernize life in the village. Paerregaard also sugggests that the 'new' Protestants are able to stand aside from the financial obligations of the Catholic *Candelaria* (Candlemass) festival, no longer a source of status within village life but for many a crippling financial burden. This economic reasoning is part of the culture of migration, contrasting with the foundation of life in the village built on reciprocity and contributions to the Catholic Church and fiestas. For the migrant, 'conversion has become a convenient way to leave the past behind' (Paerregaard 1994: 184). However, it is also clear that people (migrants and villagers) convert back to Catholicism and then convert again to another evangelical sect. Thus, religious affiliation is shifting and contingent and the increasing number of options are part of a configuration produced by the current transformations in the Latin American region as a whole.

None of this suggests an end to earlier mores and folkways that have been part of the popular cultures of urban Latin America for decades; rather, they are being re-cast for changing circumstances.

> Protestantism in Latin America may ultimately be as the Pentecostals like to say a 'consuming fire' that forges and purifies or burns and destroys. In either case it is clear that this Protestant movement is a reformation in the most literal sense of the word: a re-forming of the religious, social and political contours of contemporary Latin America.
>
> (Garrard-Burnett and Stoll 1993: 218)

Part of this re-forming, we suggest, relates to the de-centring of the social which provides an increasing multiplicity of sites in which the nation and national identities are called into being and reproduced. An evangelical church with its inclusivity and emotional appeals to sisterhood and brotherhood provides a form of horizontal integration among a community of believers addressed in an egalitarian language (although the realities of power and hierarchy may be very different). Communities of this kind invoke a correlative imaginary between the convert/believer and the Church so that they may become one and the same. The discourses in play offer a place within the parameters of the Church without reference to positionalities like 'race', class, gender or region. For those who want to leave the past behind and enter into the urban and the modern, and embrace a new identity within intelligible signposts that can be read on the basis of extant cultural competence, the evangelical churches are a seemingly safe place of belonging.

The ethnic and racial dimensions of the growing multiplicity of religious affiliations are highly relevant, part of a re-fashioning of ethnic identities in relation to political demands which cannot always be easily separated from the hegemonic power of Roman Catholicism (Rappaport 1984; Stoll 1990). In part this is related to the discourses that call forth collective subjects of struggle. The discourses from the Protestant sects privilege an account of the self and the notion of self-determination that is achieved through action and specific forms of 'technologies of the self' which are indeed disciplinary but which change the relationship between subjects and the life world in a dynamic activist way. This is not to ignore the ways in which these discourses interact in creative ways with extant commonsense and explanations of the relationship between powerful and subordinate groups, locally and nationally. But, in so far as the Protestant discourses have arrived in a specific period of change, they offer a 'new' vision and way of acting on the world which constructs an intelligible, meaningful account of the place of people in relation to change. However, given the current politics of land and ethnicities in the Latin American states, the Protestant churches are bound to engage with these issues in order to secure their positions in the localities most affected like, for example, in Chimborazo, Ecuador in the 1980s and 1990s as part of the developing indigenous movement. In this sense, therefore, the notion that the Protestant sects are part of popular cultures of resistance is relevant. However, the authoritarianism of these churches is also part of the story, comparable, suggested Lalive d'Epinay (1975) with the power relations of the *hacienda* system. Examples from Brazil and Mexico of the churches *Brazil para Cristo* and *La Iglesia de la Luz del Mundo*, founded in 1926, suggest:

> the majority of Pentecostal churches have leaders who are the chiefs, owners, *caciques* [boss, from indigenous chief], and *caudillos* [henchmen] of a religious movement that they themselves have created and transmitted from father to son according to a patrimonial or nepotistic model.
>
> (Bastian 1993: 48).

A comparative study of popular Protestant movements in Nicaragua and Guatemala in the early 1980s showed that in both cases, consistent with a growing crisis of Catholic/state relations, the Protestant sects sought new and successful ways in which to relate to the state and secure patronage (Bastian 1986). Equally, within the burgeoning democracies political parties and presidential candidates have found ways in which to secure the Protestant sects as clients. In Brazil,

> Pentecostalism has assimilated the political culture at all levels, from the electoral behaviour of leaders and members to the parliamentary behaviour of deputies. One of its slogans was the moralization of public life; but it helped to elect the most corrupt president ever and several

deputies, church leaders, social institutions and representative organs
were implicated in the Congressional corruption scandal of late 1993.
(Freston 1994: 563)

However, despite the growth of Protestantism in Brazil, it does not represent a
unity and its diversity and pluralism is 'good for democracy, whatever the
ambitions of some leaders' (Freston 1994: 564).

The relationship between religion and politics is reproduced in many parts
of the world, including the USA, where the black evangelical churches were
mobilized by Jesse Jackson for voter registration, and on the side of social
justice, at a time when many of the Protestant churches allied themselves
with the right wing of the Republican party on a platform of new moral values
– the 'moral majority'. This 'New Right' agenda, for 'the family' and against
abortion, affirmative action, citizenship rights, and so on, is racist in orientation. The discourses of what has been called the 'New Right' are the recycled
elements of a much older story. However, we need to approach the idea that
these New Right discourses are transferred to Latin America with some
caution. There is a much greater degree of autonomy among the Latin
American Protestant groups and to read these groups as facsimiles of the
North American groups is a mistake (Stoll 1990).

The contradictory story of the Protestant sects in the Latin American states
is currently emerging and the specificities within different nation-states need
to be affirmed. However, it is clear that the churches offer a vibrant space of
belonging within an increasingly fragmented world, a fragmentation to which
these very churches contribute. There is the strong possibility that the
churches and their increasing membership signal a growing disaffection with
the national project in different states in which the Roman Catholic Church
and organs of the state have played a major role. It would be interesting to
consider more explicitly the ways in which the Protestant movement offers
both trans-national identifications and local/national ones, especially among
populations who have a strong sense of their exclusion from the national
project and their place at the margins of the nation.

It is clear from this all too brief review of some of the elements of popular
culture in Latin America that, central to the configurations we have
discussed, the issues of identities and collectivities offer individual subjectives
a sense of belonging and identification, whether via the national football team
or a Pentecostalist church. Both forms of the popular offer narratives of time
and space with which people can connect emotionally and psychologically.
The importance of these spaces of belonging for national identities cannot
be underestimated because they offer, like the construction of 'the family',
mediations between local and national, big and small worlds. These mediations occur through television and, we have suggested, the importance of the
telenovela which provides national time and space and a sphere of cultural
capital which, like football at the national level, is open to all. These forms, in
this way, provide the spaces within which individuals can frame correlative
imaginaries between themselves and the imaginary community of the nation.

At the symbolic level the schisms and diversity which multiply and de-centre the social and the nation are suspended in relation to a symbolic inclusivity, in effect, a romance which is the national story. Chapter 5 takes up this story in a different way, grounding the discussion in the importance of space and place as the terrain on which 'geographies of identity' are produced and sustained.

5

NATIONALIZED PLACES?

The geographies of identity

How do the popular uses of, and movements around, the nation intervene in the constitution of national identities in Latin America? What effects have the massive migration flows had on the popular expression of senses of affiliation to place? Addressing these questions, this chapter examines citizens' relationships with variegated national places – the neighbourhood, the city, the region and so on. While Chapter 4 discussed the role of popular cultures in providing sites for the expression of national identity at the international/national interface, this chapter examines the interactions between 'local' and 'national' spheres in social practice, and in creative expressions. Non-official imaginative geographies and spatial relations with the nation are the focus, in order to provide an analysis of the 'other side' to the state project of constituting national landscapes. People hold clear ideas about how their nation is organized spatially and socially, ideas informed by popular cultures. Drawing on substantive material from around Latin America, the chapter examines the ways in which conceptions of place and an 'aesthetics of place' are articulated in discourse and practice.

These geographies of identity partially – although contradictorily – engage with, and rearticulate, the official national imaginations. Whereas official versions of nationhood draw upon localities in order to 'stand in' for the nation (Chapter 3), Andean nationals reveal a more complex and contingent relationship between 'local' places and the nation. Rather than 'standing in' in any straightforward sense for the national space, regional and local affiliations create a nuanced discursive relationship with the national, through which ideas about the exact nature of being a national and a 'local' can be expressed. To take one example, migrants to Lima's *barrios* identify themselves as Peruvians in a discourse which highlights the popular provincial inputs to national belonging, and thereby distance themselves from the metropolitan elite culture which has long stood in for Peruvian-ness officially. An imagined community of Peruvians is discursively constituted, as the parameters for defining that same community are re-negotiated.

NAMING PLACES IN LATIN AMERICA

The range of words for 'place' in Latin America is relatively wide, with the nuances of individual terms illustrating the creation of identity through the juxtaposition of boundaries and places, and even the 'playing off' of one place against another. For example, *lugar* in Spanish means a geographical place, yet *lugareño/a* refers to a rural dweller; *patria* means fatherland or nation, but *patria chica* connotes one's home town (as do the terms *tierra*, *pueblo*) (Robinson 1989). Thus names for place also significantly refer – to varying degrees – to people-in-place, populated spaces, with specific connotations about the *kinds* of communities located there. Similarly, the Quichua word for community *ayllu* refers to people in a place (Allen 1988), as does *llacta*, often glossed as town or *pueblo*. A sense of community relies upon and refers to particular places, whereby a community makes references to itself and the national society and space through a series of practices (festivals, boundary-marking, graffiti and so on). This chapter argues that the social and spatial boundaries between the 'local' and the national are dynamic, contested and multi-faceted; also, due to the mediated relations between official and popular identities, the boundaries drawn by popular citizens are not infinitely malleable, varying with 'race', gender and class positioning.

Such interlacing and complexity automatically raise questions about the supposedly fixed relationship between people and place in nation-states (see Chapter 1; cf. Balibar 1990: 337; Jacobs 1995). That such a relationship is changed through mobility or is more contingent and open to rearticulation has not attracted the scholarly attention given to citizenship relations between subjects. In general legal terms, the relationship between citizens and the space of the nation is assumed to be static: citizenship confers access to most of the national territory (except, interestingly, those areas most concerned with 'national security', such as nuclear military bases). Within the national territory subjects remain citizens through their 'natural' location; outside the territory, they are displaced temporarily (as tourists, workers abroad and so on), or the relation between people and land is broken (through war, exile, refuge).

Yet such formal conceptions of place fail to acknowledge the multi-faceted and complex *affiliations* – 'structures of feeling and attitude' (Said 1994) – to places within, and indeed beyond, the national. Although these structures of feeling are contested, socially produced and constantly negotiated, boundaries are still drawn around certain places as 'ours' or 'theirs', and distinctions between regions and nations affirmed. An understanding of the multiple relationships between the local and the national, between identities, nationhood and politics, focuses on people's uses of places and their embedded practices. By focusing on the mediations between citizens and 'their' territory, this chapter highlights a broad field of geographical imagination among citizens.[1] Such mediation holds out the potential hegemony of official national identities, yet implies too that this can shift or destabilize senses of national identity, as seen in the examples.

The chapter contains four main sections, each detailing a different dimension of the lived and imaginative geographies of national citizens. The section on 'Regionalism' looks at popular conceptions of the differences between areas *within* the nation, the identities accorded to them, and the ways in which these discursively constituted geographies inform collective identities. Rural–urban differences as well as differences between geophysical regions are examined. The section on 'Popular aesthetics of national places' deals with the ways in which imagery and language structure the 'distinctive cultural topography' (Said 1994: 61) of the nation in post-colonial Latin America. While Said traces a cultural topography through novels, our attention here is focused on oral traditions, *dichos* or sayings, and popular discourses and artefacts which express ideas about national aesthetics. The section on 'Living through the nation' examines and analyses the patterns of interaction with the national territory which are developed by subjects during the course of their life trajectories. Focusing specifically on migration experiences, military service and colonization, this section suggests that mobility around the national space may lead to profound, shifting and contradictory senses of identity and belonging with the nation. Supranational affiliations in Latin America also condition senses of national belonging; sharing Hispanic languages and a colonial history, the region's nations have several spheres through which to express common identities. The section on 'Challenging national geographies' examines how counter-hegemonic notions of place and identity are produced and re-created by popular subjects in relation to the national territory. This section argues that among indigenous groups in the Ecuadorean Amazon a process of negotiation over place, nation and identity is occurring, in which the parameters for subjective identification with the nation (as a community and a place) are highly mediated through official senses of nationhood, although the implications include the emergence of new identities unforeseen by the state. An interpretation of the interconnections between place, mobility and national identities is made, in order to formulate the idea of geographies of identity.

REGIONALISM

Strong senses of regional identity cutting across affiliation to nations are found around the world, such as Wales in Britain, Bengal in India or Eritrea in Ethiopia. Whether or not regional affiliations mobilize people into civil conflict with the state, belonging to a region calls upon a sense of community and identity which often questions and nuances national feeling in subtle and distinctive ways. In this sense then, a self-identified region may either claim a centrality for itself in the national imaginary, or feel excluded or different from the nation overall. Regional feeling may be exacerbated by struggles between different regional bourgeoisies, as occurred in Brazil and Argentina during the early republican period, if distinct 'national' agendas and associated economic and political interests are expressed. Certainly the political economic context for regionalism is noticeable in Ecuador, where the highland elite represents government, oil and agricultural interests, while the coastal elite in Guayaquil port relies upon manufacturing and export-oriented extractive and agrarian industries (Cueva 1982; Webb Vidal 1992). Developing around such economic distinctions are popular sayings and conceptualizations of the 'other' region, through which ideas about centrality and the nature of the country are expressed. Guayaquil residents say that Quiteños are dull, early-to-bed, snobbish, hierarchical, old-fashioned and conservative. The highlanders describe their port counterparts as rough, loud and like *monos* monkeys, associating them with sharp business sense, anti-clericalism and exuberance; the 'promiscuity' of the port city being perhaps attributable to its role as entry point into the nation. In this 'two city dialectic' (Whitten 1985: 35), stereotypes about people in the two cities are also stereotypes about the cities themselves. Guayaquil is described as dirty, chaotic, corrupt and full of crime, while Quito is closed up by 10 o'clock at night. While comparable stereotypes are known the world over, the point is not to reproduce here the simplistic rivalry and dichotomies which such stereotypes circulate. Rather it is to highlight the fact that these widely circulating sayings reinforce an urban-centred, metropolitan and largely 'white' elite picture of the country. Nevertheless, each place also contains social and discursive diversities which unsettle this binary.

Dichos or popular sayings create a discourse which centres on the large metropolitan areas as the significant places in the nation, thereby silencing or overwriting other urban, and especially rural, places. Discursive oppositions between Guayaquil and Quito depend upon and are grounded on an implicit, yet equally powerful, commonsense view of the *rest* of the country. Rural Andean regions of the country are pictured in a 'commonsense' way as backward, uneducated and poor, reliant upon help from outside and blocking further national development.[2] Similar Andean tropes are known in Peru and Colombia, where long-term discursive formations exist, reliant upon popular stereotypes and their selective reinforcement by official documents (Orlove 1993 on Peru; Wade 1994 on Colombia).

The Oriente region of Ecuador was for much of the early republican period off the discursive 'mental map' of metropolitan decision-makers.[3] In the late colonial era, Amazonia was not recognized as a distinctive area, although in the early republican period increasingly detailed discourses about the Amazonian region emerged. Between 1830 and 1852, the first six republican constitutions did not mention the Amazon region *per se*, that is they did not differentiate it as a distinct component of national geographies (Restrepo 1993). From the 1861 constitution, the 'Province of Oriente' appeared, although it was several more years until projects for the material (rather than imaginative) incorporation of the Amazon into the nation were realized.[4] During the course of the nineteenth century, the image of the Ecuadorean Amazon was of an inhospitable, hostile place inhabited by wild animals and 'savage tribes who worshipped the devil' (quoted in Restrepo 1993: 154). In neighbouring Peru, the Amazon region was similarly perceived until the 1930s as a place inhabited by 'aborigines' and 'primitive tribes', 'a place of mysterious allure, containing "exuberant" vegetation and untold wealth, a "zone on which the human foot has not stepped"' (Orlove 1993: 326, quoting the Peruvian geographer Paz Soldán). Given the pervasive gendering of landscapes, such primitiveness also suggested femininity – wildness and uncultured nature (Norwood and Monk 1987; Blunt and Rose 1994). In fiction, the Ecuadorean Amazon territory was feminized: the jungle represented an unredeemed female space, which infuriated male colonizers with its 'flirtatious proliferation of identities' (quoted in Sommer 1991: 269). Only later in the twentieth century has the 'distinctive cultural topography' of the Ecuadorean Oriente changed from a threatening to a promising one. The new discursive surface creating and reflecting the Oriente for the majority of Ecuadoreans (95 per cent of whom reside in other regions) is a land of promise, an unpopulated yet a resource-filled land of plenty awaiting modernization and development (cf. Chirif *et al.* 1991). The abundance of resources, imaginatively placed in the Oriente, replaces the primitive other with feminized notions of fertility, as well as with the possibility of productive development. As one of our respondents in Napo province said, 'the Amazon maintains Ecuador because it generates 67 per cent of the national budget'. However, older discursive imaginaries persist in (contradictory) popular observations that it is underpopulated, 'uncivilized' and crude.

The extent to which Ecuadoreans perceive themselves as living in a country that is defined by its Amazonian territory – which currently represents 45.8 per cent of the national total (compared with 25 per cent in the coastal plain, and 26.2 per cent in the Sierra, the remainder in the Galapagos islands) – is reflected in questionnaire responses. Asked if Ecuador was an Amazonian country, 77 per cent of respondents in highland Cotopaxi province affirmed it was. Asking *why* it was an Amazonian country elicited varying responses, the majority connected with nationalist geography and history, and with their knowledge gained from school. Overall, nearly half (49 per cent) of all Cotopaxi respondents utilized the arguments used in official discourse about the Amazon (whether in school texts and classrooms, or in government announcements). Similarly, one middle-aged, Quichua-speaking man in Coca claimed that Ecuador was Amazonian 'because [various-times President] Velasco Ibarra said it was', referring to the nationalist discourse around the Oriente from the period of the 1950s on.

The imaginative geographies of residents of the Oriente region vary from those expressed in official 'commonsense' notions of regions. Since the nineteenth-century rubber boom, indigenous groups in Amazonia have been mobile, forced to migrate rubber areas or to flee recruitment. Nowadays, the Quichua group, found mostly in Napo and Sucumbios provinces, is one of the largest indigenous groups in Ecuadorean Amazonia (CONAIE 1989; Hudelson 1987). Self-identifying indigenous subjects in Napo persistently referred to the town of Coca or the Oriente region as a whole as the most beautiful part of the country, compared with other ethnic groups who say that the Oriente is the least beautiful region. Such positive views of the local region suggest a closer affiliation to that place than among *mestizo* and white populations, both of which constitute their ethnic-national identities in relation to areas represented as more civilized.

Part of thinking about regions in social terms is that communities also think through the region-nation question in racialized terms. Racial demographics – the distribution of racialized groups around the territory – comes to stand as the signifiers of the regions themselves; 'race' is regionalized, and regions racialized. The Colombian interior is popularly imagined as a white and *mestizo* region, while coastal regions are seen as black (particularly the Pacific coast), and the Amazon region as more indigenous (Wade 1994). However, the particular conceptions of regional societies take into account specific regional characteristics as well and perceptions of distinct groups; 'there is an opposition between the Atlantic and Pacific coast, the former not so black, poor or peripheral as the latter' (Wade 1994: 64). Such diversity is nevertheless contained within the national place,

> Blacks . . . tend to be seen more easily [than indians] as Colombian citizens, albeit not typical ones nor ones that would be used to represent Colombia in most discourse about national identity.
>
> (Wade 1994: 36)

The particular history of enslavement, settlement and economic development lies behind the 'imaginative geography' which informs the Colombian national imaginary. The association of racialized (regionalized) groups with particular areas of the nation is, in Colombia as elsewhere, a result of a series of historically specific processes.

Interviews with residents of Ecuador's highland Cotopaxi province reveal clear ideas about where different ethnic-racial groups live in the nation (although sometimes they are located outside the nation!). In a diverse ethnic population (comprising 39 per cent indigenous, 21 per cent *mestizos/cholos*, 19 per cent 'whites' according to self-representation, along with 36 per cent non-attributable), there is remarkable consistency in the way those populations are imagined to be distributed, that is the regional (racialized) components of the national community. *Mestizos* are imagined to be located in the major cities (according to 47 per cent of respondents) or around the whole country (22 per cent), whereas whites are seen as predominantly urban (57 per cent of respondents). Moreover, whites are often seen as located specifically within offices! One *runa* (indigenous origin) woman said that whites were found 'in the offices', while Serrano man said, 'Ah yes, those are the ones in the offices'.[5] Blacks are seen as living in the coast, Esmeraldas and to a lesser extent Chota valley (66 per cent). Indigenous groups are identified more accurately in their attribution to rural areas and communities (51 per cent), highland areas (19 per cent), and around the whole country (11 per cent) (cf. Klump 1970; Stutzman 1974). In other words, despite a locally diverse population, comprising almost the entire range of national ethnic groups (excluding blacks), there are persistent stereotypes about where different groups are located in the nation. Such racialized imaginative geographies circulate largely within popular culture, owing their details only slightly to official conceptions of how the national community is organized.[6] Although there is some re-representation of the country as containing picturesque indigenous communities in the Sierra and the Oriente, such official re-mirroring draws heavily from popular imaginative geographies.

Although often racialized, the relations between nation, place and 'race' are historically variable, shifting with the political and social context within which people make affiliations to place and the communities found there. The identification of Bolivian highland indigenous groups reveals profound shifts around different 'scales' of affiliation through two centuries. The evolving identities of the Aymara group illustrate a high degree of fluidity in the interface between 'race', place and nation. Between the late eighteenth century and the present day, wide regional ethnic identification among the indigenous population in the *altiplano* (highland) was replaced by parish-wide or local identities, based on birthplace (Albó 1987; cf. Platt 1993).[7] Although the denomination *campesino* offered some possibility for a wide regional identification during the mid-twentieth century (in addition to its class connotations), it failed to erase the implications of indianness, inferiority and marginality which adhered to rural *altiplano* dwellers. In this context, a culturally oriented identity based on language and cultural difference arose in the 1970s, mobilizing a wide range of social subjects whose identity as 'Aymara' was successfully interpellated, particularly by the nationwide Katarista peasant movement. Nevertheless, this cultural identity – with its corresponding socio-spatial spread – is underlain by and co-exists alongside an *Andean* identity, into which diverse local identities are articulated. The co-existence of identities at various spatial scales, each with its own particular 'history' and 'geography', thus breaks down any idea of an essential 'Aymara' identity with given social and spatial boundaries and unchanging trajectory (Figure 5.1).

One feeling of affiliation to place found in Latin America is that expressed in the term *patria chica*, literally meaning small homeland. Referring to a migrant's place of origin or a place where one feels at home, the notion of *patria chica* raises immediately the issue of overlapping affiliations to place, the *patria* and the place of origin. While not having such a widespread usage in Ecuador as in Peru or Colombia, popular naming of *patria chicas* raises some interesting issues about belonging and affiliation in light of other facets of identity. In Ecuador's Napo province, a young 'white' man identified his *patria chica* to be Quito and Galapagos, an interesting conflation of two sites (cultural/natural, central/peripheral) which feature prominently in official representations of the nation.[8] Among Cotopaxi residents interviewed (over half of whom spoke Quichua), nearly a third identified a *patria chica*, largely in terms of their birthplace and familial affiliations. Yet a quarter of those affiliating themselves with a *patria chica* in Cotopaxi claimed a sense of belonging with places (either on the Costa or elsewhere in the Sierra) they had migrated to and lived in during the course of their lives.

The above examples from Colombia, Bolivia and Ecuador evidence people's understanding of regions and difference *within* the nation, which revise popular senses of place and where different communities 'belong' within the nation. People can feel 'Peruvian' or 'Colombian', yet the ways in which they think about that identification is in relation to non-local places within the nation, organized in terms of regions and racialized populations. The material also suggests that the places through which people imagine their communities and their affiliations to place are not fixed, but are liable to shifts resulting from socio-political mobilization. The Aymara example showed how what appears as an essential social identity is expressed through malleable and spatially contingent communities and places, constantly transformed through Bolivian history. The next section goes on to examine how these cross-cutting identities are also bound up with non-verbal representations and aesthetics.

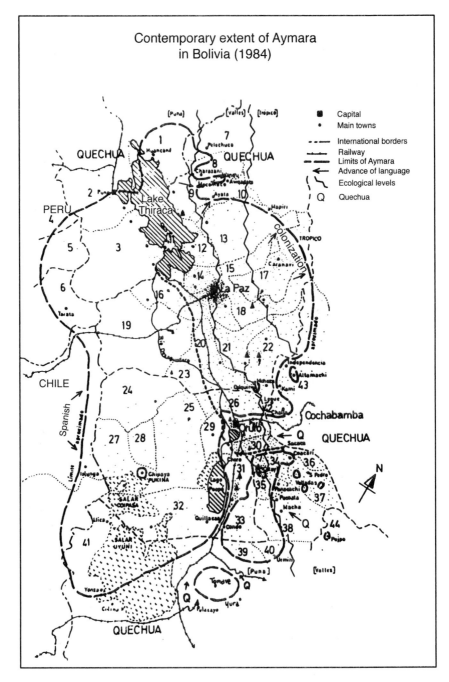

Figure 5.1 The socio-spatial extent of Aymara identity in 1984, showing the dynamic
interaction of Aymara, Quechua and Spanish languages

Source: adapted from Albó (1987).

POPULAR AESTHETICS OF NATIONAL PLACES

'The connection between the imaginative geography of landscape and the imagined community of the nation' (Daniels 1993: 243) is one which has largely been explored in terms of landscape painting in metropolitan countries. However, in non-metropolitan countries – in both 'high' and 'popular' culture – the connection between aesthetics, landscape views and national identities is widely appreciated (Rowe and Schelling 1991; Canclini 1993). This section explores the connections between aesthetic representations of the national space, national identities, and the role of material culture in providing national icons. In Latin America, just as elsewhere in the world, the production, circulation and consumption of 'national' aesthetic products results from complex interactions involving relations of hegemony and contestation between 'popular' cultures, the state and nationalist ideologies (Canclini 1993; see also Chapter 4).

In Latin America, in common with English landscape paintings and national identities, there is seldom a secure or enduring consensus about which particular landscapes and legends epitomize the nation (Daniels 1993: 4). This is certainly true for Ecuador where people have very varying ideas about national landscapes, depending on their class, 'race', location and age. One arena for the circulation and consumption of aesthetic representations of nationhood is the Casa de Cultura in Quito, visited by thousands of school children and adult citizens every year. One of the largest rooms is the nineteenth-century room containing portraits of the Independence hero Simón Bolívar, as well as past urban landscapes, former national presidents in interior scenes and allegorical views of '*América*'. The landscapes hanging in the gallery depict specific places and events in Quito reaffirming – through representations of urban squares and historical 'national' events occurring in the capital – the centrality of urban centres in the national narrative.[9] 'National' history is pictured in metropolitan, urban and predominantly 'white' social spaces, and the place of Quito in this national museum 'stands in' for Ecuador. Moreover, the conflation of metropolis with nation reiterates the role of the city as the departure point for the colonizing/development project of progress (Jacobs 1996). Similarly the interior portraits of past presidents refer to the centres of decision-making and domestic spaces of the central elite, in their robes and with various badges of office, such as sashes and documents.

At the same time as these aesthetics are presented through public display, other representations of 'Ecuador' and its people are produced and deployed. One of the most widespread artefacts are small paintings of rural, particularly Andean, settings, known as *pinturas de tigua* (Tigua paintings). Consumed by tourists as well as Ecuadoreans and displayed in numerous streets and parks, the *pinturas de tigua* provide an alternative aesthetics of national landscapes (Figures 5.2 and 5.3). Drawing upon patterns of life based on subsistence-oriented mixed-farming economies, the *pinturas* use a naive style with little or no perspective in addition to bright colours to show figures, animals, rural houses, Andean settlements and landscape features. Acrylic paint is used on leather stretched over wooden frames to re-present 'national' spaces, from the perspective of the indigenous peasantry of the Sierra. Domestic scenes (such

Figure 5.2 Pintura de tigua from Ecuador, showing an Andean rural landscape

as the birth in Figure 5.3) represent adobe houses and the networks of family and community which permit daily and generational reproduction. Placing themselves firmly within the national imagined community, the producers of *pinturas de tigua* simultaneously re-centre the nation – in the Andes – and 'nationalize' an area which has long been marginalized and not generally imagined as part of the nation.

Drawing upon distinct ideas of environment, morality and politics, popular representations of the Andean mountains challenge not only the centrality of Quito in 'national' imaginations, but also the socio-economic hierarchies which flow from that. Representations of mountain peaks, and indeed the mountains themselves, draw upon contemporary Andean cosmologies which challenge and limit secular state power (e.g. Guerrero Arias 1993; Silverblatt 1988; Radcliffe 1990a). Juxtaposition of national flags and mountain peaks in the *pinturas de tigua* draw upon complex discourses that specify both the interlinkages and distances between peasants and nation. Linkages are exemplified by state titles and judicial systems which guarantee land-holdings and peasant national identities, but are conditioned by peasant distinctions between state political powers and the lifegiving powers of the mountain *apu* (mountain deity), and by peasant scepticism about the state's ability to deliver development to rural areas.

The complex, ambiguous relationship between peasant and nation from the peasants' perspective of *pinturas de tigua* is contextualized by official uses of these artefacts. *Pinturas* are widely promoted by the Ecuadorean state in the tourist business, as well as reproduced in postcards and brochures, thereby in part appropriated by the state. Yet the widespread display and consumption of

Figure 5.3 Pintura de tigua, showing a domestic interior after a birth.

the *pinturas* profoundly reconfigures the 'national' landscapes through which Ecuadoreans imagine their place. Especially with the consumption of *pinturas* by foreigners, the Andean-focused aesthetics of national space are reflected back into Ecuador with subtle effects on identifications. While Canclini (1993) suggests this process is one of 'hegemony allied with subalternity' to produce hybrid cultures, the *pintura* goes beyond this in its reconfiguration of both hegemony and subalternity through the re-territorialization and re-mirroring of the national aesthetics of place.

Elsewhere in Latin America, popular practices are used to renegotiate around official views of the nation, and its representation. The public consumption of a red-and-white striped alcoholic drink in the Andean, largely *mestizo* provincial town of Sicuani, southern Peru, draws upon an aesthetic (as well as economic and political) sense of identity (Orlove 1986). The ingredients chosen to colour the (otherwise colourless) sugarcane alcohol *trago* are trade goods, rare elsewhere in the southern Andes but encountered in Sicuani due to its position at the hub of widespread trading and marketing networks extending to the southern highlands and the coast. The striped colouring of the *trago* mirrors the colours of the Peruvian national flag – red, white, red – and positions Sicuani firmly within the nation, recognizing its legimacy. However, by using local-regional goods to make the image of the national flag, this apparently patriotic practice is linked with demands for regional autonomy and regional capital status. As in the Ecuadorean *pinturas*, the visual representation of the national flag reflects and enlarges upon a discourse of *local* belonging within a (thereby qualified) sense of national identity.

Popular subjects express detailed aesthetic judgements about different places in the nation, outside the context of formal occasions and particular representational forms. Ecuadorean interviews demonstrate evidence for people's aesthetic discrimination, and imaginative picturing of different national locations. Moreover, very few respondents declined to name one specific pleasing place in Ecuador, although they were more reluctant to name unpleasant places.[10] For popular subjects of diverse ethnic identifications in highland Cotopaxi province, the most beautiful Ecuadorean places were predominantly in the Sierra, reflecting their own experience as well as official national images (cf. Chapter 3). Over half (51 per cent) identified towns or natural features in the Sierra as the most beautiful national place. Quito was the most aesthetically pleasing place for one-fifth of respondents, reflecting its (official/popular) centrality in national imaginations, while 11 per cent identified other towns (Ibarra, Imbambura, Baños, Chilla Grande peasant community, and 'the city' generally). Natural features (including the Galapagos islands, Chimborazo volcano, beaches, the Sierra landscape) were highlighted by nearly half, choices mediated by the externally oriented marketing of Ecuador for international tourists. Despite much official attention paid to the Mitad del Mundo site, only one person mentioned the monument as the most beautiful national place. Discrimination between national features on aesthetic criteria significantly differed from people's sense of what represented Ecuador in the world. Here again natural and landscape features (including the Galapagos, Chimborazo and Cotopaxi) figured prominently, yet nationalistic sites (Mitad del Mundo, the Equator line and other tourist sites), and objects (flag, shield, textiles) became more important.

LIVING THROUGH THE NATION

While knowledge and images of the nation are constituted through verbal and visual representations, the material and experiential procession of citizens through the national space is another (more frequently neglected) aspect of the formation of national belonging. National senses of belonging are dynamic spatial processes, developing over space(-time) and through specific places, through such practices as religious pilgrimages, migration for work and colonization, and military service. A nation has inherently, a dynamic relation with place, as well as comprising a history and a society. While other commentators have rightly pointed to the centrality of territory and internal spatial organization of nations (often equated with the division of territory into administrative areas: e.g. Williams and Smith 1983), the full implications of taking the nation as a geographical imagination and practices have not been fully taken on board.

Anderson's (1991) book on national imagined communities relies upon a (largely unacknowledged) geographical imagination which permits him to link themes of space, mobility and the nation. First, he details the movement of national leaders of colonial states in South-east Asia, discussing how these leaders, as school children and then as university students and administrators, moved through urban hierarchies and institutional networks created by French and British colonial rule in the region. In the pre-independence era, individuals came to experience, represent and most significantly *embody* a 'nation' through their knowledge and education gained in various parts of the territory (see Chapter 6). Institutions, geographies and life-history came together to offer the possibility of what we might term a 'template' for an alternative, a post-colonial independent nation. A life trajectory was mapped out on to a geographical-social 'national' grid, and subjectivities (ideologies of 'nationalism') were constituted through experiences in diverse places (first colonial, then national). Anderson focuses predominantly on a small elite of mobile 'native' intellectuals, leaders of nationalist movements, anti-colonial ideologues, whose experience was interpreted and then mediated through colonial and post-colonial institutions. The rationalization of their trajectories served a particular function during the foundation and consolidation of a post-colonial nation.

By contrast, the experiences of labourers, domestic workers, migrants and conscripts – to name just a few – are not mediated at such levels, yet reveal other dimensions to identity formation and mobility. In contemporary Latin America the experience of migration is a democratic one, in which large numbers expect to participate as equals and in which ideas of belonging, nation and position are all self-conscious components in discourses around the urban. Migration draws upon notions of individual and family advancement, engagement with national institutions (schools, hospitals, public services, state administrations) and ideas of progress and development, and thereby occurs within a highly loaded arena. Where massive urbanward migration occurred, especially after the 1940s in many Latin American countries, migrants' socio-political situation and experiences conditioned the engagement of popular identities with issues of territoriality and belonging. While movement was not sufficient for the formation of national consciousness, the socio-economic changes attendant upon migration were salient in terms of *how* (in which institutional and social contexts) and *where* (in urban/rural areas, particular cities and/or regions) identities with place were constituted and experienced. With urbanization signifying progress in the developmentalist discourses of nationhood, migrants potentially accessed both cultural capital and social resources through arrival in the city. The imaginary in which the city epitomizes the nation's progress – as shown in the above discussion of paintings – is reinforced materially by the allocation of resources and population movements.

In Latin America's massive urbanward flows of populations, people's experiences of place (and associated cultures, societies and economies) were radically transformed, as were national imaginaries. Although the southern cone countries of Argentina, Chile and Uruguay were highly urbanized early in the twentieth century, other countries saw rapid and permanent urbanward movements of population between the 1950s and 1990s, largely towards the huge metropolitan centres. While 42 per cent of the region's population was urban in 1950, this percentage had risen to 72 per cent in 1990, resulting from urban annual growth rates averaging nearly 4 per cent over those decades. Not surprisingly, such demographic changes brought in train a series of wide-reaching social and economic transformations, not least in the explosion of marginal *barrios* and poor neighbourhoods on the outskirts of major cities (the downside to discourse of urban progress), the development of informal economic activities, and the creation of new synthetic popular urban cultures. In terms of national identities, the move to the city was significant in many ways. Migrants were often aspiring to greater educational opportunities for themselves and their children, which brought them into greater contact with nationalist schooling procedures throughout the educational system (Chapter 3). As women were predominant in rural–urban migration flows in Latin America through much of the mid-century, and gained more education (especially among younger age groups), this access to urban areas and attendant services was particularly significant for female national identities (Chapter 6). Moreover, the insertion of migrants and their families into the urban political process often appeared to transform their sense of national belonging and articulation of 'community'. Nevertheless, migration and subsequent socio-cultural transformations did not necessarily mean an end to belonging to other places; provincial and village loyalties remained and were even strengthened among certain populations.

These themes can be explored in the case of Peru, a country where migration resulted in a doubling of the urban population between 1950 and 1990 (36 per cent to 70 per cent urban), and whose capital Lima grew to over 6 million inhabitants during the same period. Working in the *barrio* of San Martín de Porres, to the north-west of Lima, Degregori *et al.* (1986) traced the transformations in migrants' senses of place, origin and identity from the initial period of settlement through to the 'second generation' of migrants' adult children. In interviews with migrant families, a multiplicity of 'overlapping' senses of locale and belonging emerged. Whereas before migration identities were expressed in terms of affiliation with a village or town, early migrants to Lima became aware of the diversity of their neighbours' origins and hence the co-existence in homogeneous time of many provincial towns and villages. As one respondent, Mateo, explained,

> I have friends from various places, as many from the coast as the Sierra. I have quite a few contacts, I don't find much difference – when I meet three people from Cusco, Arequipa and Ayacucho, for me they're equal/the same, because I feel like them.
>
> (Degregori *et al.* 1986: 163)

Identifying as Peruvian was the discursive response to a shock of recognition with other migrants: 'I identify myself as a Peruvian that I am', said Pedro, a migrant. Quotidian life in the *barrio* opened out a sense of national belonging, allowing migrants to create discursively 'the image of their communion' made up of provincials like themselves (Anderson 1991: 6).

Yet national identities were also expressed in relation to a plurality of cultures which together make up the 'national': being Peruvian did not cause the automatic loss of diverse regional customs and 'traditions'. Migrants' national identity drew upon their self-conscious proud knowledge of, and participation in, diverse cultural practices associated with subnational regions and not with the whole country. As migrant Elsa explained,

The customs of one are the customs of all; I'm not a *serrana* [from the Sierra] yet I dance *huaynos* [a dance associated with Andean peasants]; those from Sierra eat food from the North. So it's like that for us, it's the same [*igual*] – we don't say 'I like this because it's from there or here'.

(Degregori *et al.* 1986: 163)

However, recognizing diversity within the national does not guarantee the reproduction of all 'local-regional' practices, as music, dance and song reveal (for other examples of these urban cultural reconversions, see Vila 1991). Upon migration to the city, Andean migrants often become reluctant to use Quechua with their children or other *serrano* migrants, while they also often give up 'typical' Andean clothing styles. By contrast, the general acceptability of Andean music was framed and legitimized by state legislation passed under the corporatist Velasco military regime, which from 1975 required all radio stations to dedicate a minimum of 7.5 per cent airtime to 'folkloric music' (Turino 1991; 1990; Llorens 1991). Defined as music 'born directly from the traditions and customs of the people . . . and especially from the rural zones', folkloric music took an average of 15.2 per cent of Lima's airwaves by 1987 (Turino 1991: 274–5). So in contrast with Guatemala where *traje* (clothing) has been culturally reconverted by the state in a national cultural practice (Chapter 1; cf. Sommers 1991), in Peru it is music, through which the 'national' is re-imagined and reconfigured.

Regardless of changing affiliations to place, first-generation Lima migrants retained links to their areas of origin, via the numerous regional and provincial associations in the capital (Altamirano 1980). By contrast, the youngest *barrio* dwellers often have not visited their parents' birthplace (Degregori *et al.* 1986: 255). Rather than being linked to a specific province, Lima's migrants become identified by others and identify themselves as either *costeño/as* (coastal dwellers) or *serrano/as*, 'members of one of the two great poles of the historic dichotomy which splits the country' (Degregori *et al.* 1986: 255). Nevertheless, the youngest migrant groups have a strong sense of the equal value of *serrano* and *costeño* as broad social cateogries, less tied to the discursive hierarchies that placed *costa* over *sierra*. In this sense, their vision is more 'national', identifying broad regional differences while denying demeaning stereotypes. However, Degregori, Blondet and Lynch question whether this equality is tied specifically to the migrants' location in Lima, where the stigma of regional origin can indeed be minimized by participation in shared activities and spaces.[11]

Yet such historically persistent identities of *costeño/serrano* are contingent, liable to be set aside in favour of an encompassing *provincial* identity when faced with the racist, centralist and seigneurial society of Lima. As one respondent Luisa explained,

The provincial for me is . . . well, it depends, because also there are lazy provincials, but the provincials are good workers, they don't say 'well, now I'm tired, now I don't want to . . . and my hours' and so on. No, they go on, whereas the Limeño is limited.

(Degregori *et al.* 1986: 270)

Provincial identity – when positioned *vis-à-vis* an imagined Limeño community – is hardworking and uncomplaining; other terms of reference include the *serranos'* honesty and greater willingness to struggle. Containing elements of class identity, the positive provincial identity draws also upon a dynamic popular culture (partly provincial but mostly newly synthesized), an alternative to elite high culture claimed by the Limeño bourgeoisie. As Hugo, a 'second generation' migrant explains, 'I prefer to feel more Peruvian than Limeño' (Degregori *et al.* 1986: 163). Being Peruvian rather than Limeño, as the last respondent says, claims a national identity and simultaneously qualifies it: 'I am Peruvian but not in the old sense of Peruvian, meaning Limeño' is the subtext here. In addition to highlighting other affiliations to place alongside national identities, the responses reveal the negotiations around the 'national' in terms of class and location.

> [The migrants' children] prefer to define themselves – as do a sector of the older migrants – as *Peruvians* of San Martín de Porres, inhabitants of a Lima divested of its creole and colonial attributes to convert itself into a battleground between a predominant tendency to transnationalization and another to 'Andeanization' or 'cholification'.
>
> (Degregori *et al.* 1986: 26, emphasis in original)

The Lima case provides an excellent illustration of the expression of (multi-faceted) identities through references to a diversity of places simultaneously. Moreover, identity is expressed by these migrants in relation to a variety of spatial scales, according to context and biography. Migrants do not replace a provincial or village identity with a fixed national identity, but retain and renegotiate from their multiple positions. Migrants constantly re-express their identities in terms of the neighbourhood, Lima, provincial origin, *serrano/costeño* and Peru, according to the discursive and relational elements associated with each. Each geographical affiliation with its own sense of belonging allows the migrants to intervene in debates about Peruvian nationhood.

Another movement in which large numbers of citizens participate (although limited to certain gender and age groups) is conscription and military service, a masculine civilian obligation in most Latin American states since the early twentieth century (introduced in Chile in 1900, Ecuador in 1902, Brazil in 1916). Military lifestyles and 'traditional' masculinities have often been closely associated in representations and senses of self in Latin America related to modernist discourses of order, hygiene, education and moral duty. While professional armed forces draw upon middle-class, upwardly mobile, often *mestizo* men in most countries, military service recruits more socially and racially diverse groups for short periods of time; in Brazil, 50 per cent of navy crews are black (Rouquié 1987: 97, 64). The gulf between highly educated, socially segregated, professional officers and the conscripts can be large, especially in the most multi-ethnic societies (Duncan 1991; Rouquié 1987). Officers may feel superior to the conscripts, as in Bolivia where enlisted men are predominantly 'indian', and officers are white. In Ecuador, such racialized difference is not so prominent, with *mestizo* officers coming predominantly from small towns and rural areas, mostly in the Sierra (Hurtado 1977: 248).

Voluntary recruitment has rarely been used except in Uruguay, yet contemporary obligatory military service has often been evaded, just as it was in the nineteenth century (Rouquié 1987: 94). Currently, the Ecuadorean constitution requires men over 19 years of age to give 12 months' service in one of the camps around the country (Molina Flores 1994). As the majority of conscripts are not professionals, the armed forces aim to provide 'an opportunity to receive a basic education and training in jobs which will serve once [the conscript] is reintegrated into civilian life' (Molina Flores 1994: 141). Generally such education is predicated upon an imaginative community speaking Spanish and knowing the nation. For example, the Spanish literacy programme introduced into the Guatemalan military dealt with the 14 per cent of conscripts who spoke no Spanish, and the two-thirds who were illiterate (Rouquié 1987: 97). Ecuadorean military education is highly patriotic, with recruits exposed to more intense and frequent veneration of the national symbols – the flag, anthem, arms as well as its territory and history – than school children under national curriculum requirements (Hurtado 1977; cf. Chapter 3).

It is noteworthy, however, that only a small percentage of the male population interviewed in Ecuador had done military service, with figures around 16 per cent in Cotopaxi for example, mostly men over 45 years.[12] Thus although military masculinities are commonly supposed to express national identities in Latin America, in practice it is just one masculinity among several (Chapter 6), and one way of experiencing different regions of the country. In our interviews, it became clear among provincial indigenous and *mestizo* populations that military service was perceived as something to be avoided. Some villagers prefer to pay a fine (equivalent to US$164) freeing them from this obligation, and others risk not paying it (and being forced into a military lorry on market days).[13]

Yet for men who do participate in military service, the effects of being moved into another town or region with a diverse group of co-nationals is a profoundly transformative experience. For one former Ecuadorean conscript now living in a lowland town, the experience did not take him away from the Oriente region (he was posted to Shell Mera), yet it inspired in him a love for the entire country and for the framework of formal citizenship. In contrast to other respondents in Napo province, Eloy claimed that he had no ethnic identity or *patria chica*, but loved the whole country and that citizenship rested on knowledge, indicative of the emphasis placed on continuous education in the military. Ecuadorean indigenous commentators concede that military service succeeds in integrating indigenous groups into the nation. However, racialized relations between conscripts and officers may problematize the imagination of a shared community, where, as in Guatemala, over half the officers felt that indians were inferior. In such circumstances, conscripts are more likely to retain regional, ethnic or village affiliations rather than identify with a national community which discriminates so directly against them.

Enhancing masculinities is also a key component of military service, with conscripts claiming virility and *machismo* (which is perceived as positive), in opposition to the feminized men evading the draft. Cultural expressions of masculinity change in Ecuadorean highland villages with ex-conscripts identifying themselves as *'buen varón'* (real men), *'un supermacho'*, who *'tienen huevos'* (have balls) in contrast with the *'cocineros'* (cooks) who stay behind. Integrating subjects into national identities through military service is then about enhancing masculinities and deepening gender difference, as well as creating a 'national' masculinity through which to claim difference from neighbouring countries (cf. Enloe 1989; see also Chapter 6).

In summary, military service provides perhaps the most structured and 'patriotic' means through which male citizens can take part in movement around their country, meeting co-nationals and re-imagining their community. Nevertheless, the limited extent to which men, especially younger men, are now fulfilling their civil obligation of military service suggests that this avenue for rearticulating affiliations with places and people is not as significant as work- or family-related spontaneous migration in forming national identities.

Oriente colonization and senses of place

As noted on pp. 110–11, a switch in the Oriente's representation pictured a place with resources which promised and underpinned potential development. Such re-presentation entailed the state-promoted move of 'appropriate' populations into the region, in order to exploit the resources and guarantee production (Chiriboga 1988). Although only one of numerous groups in the Oriente (also including missionaries – Protestant and Catholic, oil company workers, non-governmental organizations, merchants, the military and bureaucrats), *colonos* (colonizers) are pictured as having a unique, specifically national-developmentalist, relation with the region. The nationalist emphasis on colonization as an agent of development in frontier regions (as in other countries: Townsend 1995) means that the Ecuadorean colonizers have a particularly significant role in the constitution of national identities. Creating a 'new national cultural space' (Little 1994), colonizers from the coastal and Andean provinces have been encouraged to settle in areas previously inhabited by indigenous groups, missions and oil companies. The founding in 1958 of the Instituto Ecuatoriano de Reforma Agraria y Colonización (IERAC: National Institute of Colonization) and the *Banco Nacional de Fomento* (National Investment Bank) to oversee colonization prompted the movement of thousands of generally small landowners or landless peasants (Uquillas 1984). Moving to the coast in small numbers, and later to Amazonia, colonizers have been seen as the agents through which Amazonia can be brought into national development, and realize its nationalist goals (Ruiz 1994).

That the colonizers feel themselves to be agents for the realization of nationhood is borne out by interviews in the settlement of Tarapoa, in Ecuador's northern Oriente: 'before the colonizers came, this wasn't Ecuador, this wasn't Colombia, it wasn't anything', said one colonizer explaining the reason for settlement (quoted in Little 1994). Tarapoa identities are constructed from local, national and international practices. Being a secular society, the *colonos* form a community through social dances, as well as football and beauty contests.[14] In dances on a newly constructed multi-purpose surface, colonizers bring in international and continental culture, via the diverse music such as rock, *cumbia, merengue, salsa* and *son*. The Tarapoa beauty contests allow for the staged juxtaposition of regional and national elements, through the pacific co-existence of ethnic diversity (also Chapter 3), through the reconversion of international contests broadcast by satellite. For men, weekend drinking sessions, football and cockfights offer opportunities for sociality in place.

Despite the diverse provincial origin of colonizers (thirteen different birthplaces, plus two from Colombia out of a population of 3,000), *colonos* feel themselves to be Ecuadoreans. 'I was born here, I will die here. What else can I do? My flesh is Ecuadorean', according to one respondent (Little 1994: 558); one of our Napo respondents invoked a similar embodiment, saying that 'Ecuadorean blood runs through my veins'. Yet this embodied national identity is qualified by a general sense that the nation has marginalized *colonos* and frustrated their goal of development (which they presumed was shared with the state). Although engaged very fully with the state's developmentalist rhetoric and agenda, colonizers feel their work on clearing forest deserves recognition in the form of land titles, which are not forthcoming. Acting as a workforce for the state, rather than as citizens with land titles, *colonos* lie in an ambivalent and marginal relation to national place. Attempting to carve out a livelihood for themselves in a new ecological and social environment, conflicts often arise between colonizers and indigenous groups particularly over land claims (Ortiz 1987; Albornoz 1971).

CHALLENGING NATIONAL GEOGRAPHIES

[The missionaries] considered us devil-worshippers, idolators. We were taught to respect authorities which were not Shuar, and taught that we were part of a different homeland [*patria*] so we couldn't visit our brothers who stayed in Peruvian territory, converting us to foreigners in our own territory.

(CONAIE 1989: 9)

[Amazonia] is neither the salvation of the country nor the lung of the world, but a region which it is necessary to know very well in order to search for development alternatives for the Amazonian peoples.
(Valerio Grefa, President of CONFENAIE, Ecuador,
in Grefa 1993: 418)

In the face of frequently incompatible official notions of territory and community, Amazonian indigenous nationalities have been actively involved in creating their own senses of place and nation. Faced with numbers of incoming settlers from the Andes and multi-national oil companies encouraged by the central state, the response of indigenous federations has been to attempt to clarify their own 'places', namely the large territories upon which their livelihoods depend. Particularly given the discursive construction of the Amazonian region in Ecuadorean national self-representations, the indigenous groups in the Oriente have countered prevailing images of the area and its inhabitants, in order to create a more realistic imaginative geography. Through their confederations, indigenous communities have argued that the Oriente region is *not* empty but widely inhabited, by groups who already use rainforest resources (CONAIE 1989; Mundo Shuar 1983).

Moreover, indigenous confederations argue that the nationalist 'freight' carried by the region – its promise for future national development, as a site for national realization – is not justified. Rather the Oriente is re-represented by indigenous federations as another part of the country, with its own development problems and dynamics (Grefa 1993: 418). Yet official and non-local imaginative geographies have influenced indigenous Oriente populations sufficiently for them to feel alienated in part from the region, and marginalized. As one self-identifying indigenous man told us, 'I don't feel part of the Ecuadorean nation, because we don't have permanence, consistency, in one place, and we are not an integral part of the nation'.

In addition to migration, cartography and mapping are important for providing a screen on to which national imaginative geographies can be projected. The ways in which cartographic practices and visual representations enhanced and 'grounded' colonial and post-colonial identities is by now relatively well documented. Cartography historically produced images which claimed scientific status and objectivity, produced by certain interest groups to convey their (often privileged) view of the world (Harley 1988). While geographers have highlighted the power-driven use of cartography and mapping (usually) by elite interest groups, particularly under colonialism, the current section explores the complexities and slippages around mapping and map-makers (also Jacobs 1995). A more open interpretation of mapping and maps is crucial for understanding cartography's use by non-official Ecuadorean organizations, who have developed map-making procedures to represent alternative discourses and representations of the nation and its territory. The Ecuadorean state has availed itself of cartography as a dimension of the power-knowledge of the nation, its territory and its sovereignty. Rather than taking mapping's inherent power-filled agenda as a premise, the case study from Ecuador's Oriente region suggests that cartography represents a practice through which popular and official senses of place and identity interface. In other words, the process and production of mapping are not inherently restricted to those in power; map-making can embody resistance and alternative agendas to those associated with traditional imperial or state cartographies (e.g. Cant and Pawson 1992).

Since the mid-1980s, indigenous confederations particularly in Amazonian Ecuador have been engaged in mapping projects as the basis for political negotiation with the state over issues of land-rights, social organization and multi-culturalism. The setting up of cartographic teams by various self-identified indigenous unions illustrates a process of taking back the power/knowledge of cartography from the nation-state and using its techniques in underpinning alternative political projects. Describing these cartographic efforts however reveals the impossibility of setting up stable and fixed oppositions between 'official' and 'indigenous' versions of national place, and the role of mapping in them. Thus while Loomba rightly points to the need to recover 'alternative versions of nationalisms' (Loomba 1993: 306), the contemporary Ecuadorean indigenous topographers, engaging in Spanish-language legal procedures, illustrate the interlacing and blurred edges of official and 'non'-official geographies of identities. In other words, there can be no simple identification of the 'marginal', in this case the indigenous topographers, as the privileged site of resistance; the binary of incommensurate difference assumed to underlie notions of resistance breaks down. Maps are used by indigenous groups to gain a position which, while not unmarginal, engages with the mapping process and offers the possibility for new identities.

Since its foundation, the Confederación de Nacionalidades Indígenas de la Amazonía Ecuatoriana (CONFENAIE: Confederation of Indigenous Nationalities of the Ecuadorean Amazon) has seen the recuperation and defence of indigenous territories as one of its major mandates from members (Serrano 1994: 437), as have other regional federations such as the Organización de Pueblos Indígenas de Pastaza (OPIP: Organization of Indigenous Peoples of Pastaza Province). By making visual representations of indigenous nationalities' land, the confederations represent political and ideological interventions in state-defined land allocation and titling procedures. The popular geographies exemplified by indigenous confederations' maps refer implicitly to – and are contextualized by – a series of social, political and administrative agendas for change, beyond the land issue *per se*. Currently, some 60 per cent of indigenous nationalities are in the process of recovering their land. However, some two-fifths of indigenous lands are under state control, located in reserves, national parks and forests protected by the Ministry of Agriculture (Grefa 1993: 417). Indigenous lands have long been settled by incoming Ecuadoreans, usually indigenous and *mestizo* peasants from the highlands, encouraged by numerous state development projects.

In order to receive official title to lands, indigenous federations negotiate with the state using the judicial system as well as legalistic practices and discourses, presenting petitions to government and lobbying politicians and trans-national oil companies. Within this Spanish-language process, the creation of visual representations of landclaims is important. Topographic teams formed by the confederation create cartographic representations of the landclaims, in order to delineate and name the extent and nature of their territorial claims, as well as to make a cultural product which counters the widespread view that the Amazon is unpopulated. The maps thus serve multiple functions – affirming local knowledges of boundaries and settlements, giving appropriate visual form to landclaims, and reaffirming the geographies of identity which motivate indigenous federations in their actions. Moreover, marking the territory reminds younger villagers of the borders and reinforces the mnemonic function of border-marking festivals in certain Quichua and Shiwiar settlements.[15]

Map-makers in the indigenous federations are thus key actors in the marking-out and realization of indigenous projects. Topographic teams, formed by various indigenous federations such as the Shuar Federation and the OPIP since the mid-1980s, have mapped out the Amazon territories of indigenous villages. To give an example, the OPIP federation is a multi-ethnic organization including Quichua (the majority), Achuaras, Záparo and Shiwiar nationalities. Trained by a German development agency technician in 1984 during a short course at the headquarters of the CONFENAIE, the OPIP topographic team comprises two or three indigenous men who work for other indigenous organizations as required (also Chirif *et al.* 1991).[16] The topographers tend to work on a temporary basis, as and when required by villagers, and draw upon the knowledge of boundaries held by villagers themselves. Moving through the landscape in teams made up of the topographers and local people who know the settlement frontiers, the topographic team marks the recognized boundaries on a map, as well as clearing vegetation in a three-metre wide strip on the ground. To confirm borders, the teams put up zinc plaques saying 'Quichua territory, so-and-so village, Made by OPIP', and also paint the surrounding trees with kilometre measurements. When borders lie between different indigenous groups, such as between Quichua and Shiwiar, representatives of both groups attend the mapping and marking process in order to gain consensus.

In the most significant piece of work done by the OPIP topographic team, a map was utilized to lobby the state for a major landclaim. During the mapping process, nineteen communities were visited in the boundary-marking work, as well as the colonizers near the Pastaza provincial capital of Puyo. The boundary-marking process – or *autolinderación* (self-boundary-marking) as it is known – resulted in a map which lay claim to a multi-ethnic territory of some 1.1 million hectares, one of the most coherent proposals for territorial organization in the continent at the time (Chirif *et al.* 1991: 58; Radcliffe 1996a). As only two settlements in this area had official land titles, the proposal to recognize all Quichua, Shiwiar and Achuar lands in the province was a high priority for OPIP (Figure 5.4). The proposal, based around the topographic map, was presented to the then-President Borja during the indigenous Uprising in 1990 (Declaración de Quito 1990; Zamosc 1994; Karakras 1992), and in 1993 the Pastaza land rights were recognized.

The confederations' success in creating a space for their identity – a space represented visually (on maps and land titles) and defended politically – cannot be denied. Moreover the legal procedure for landclaims has been changed as a result of the confederations' persistence. Originally the Ecuadorean state conceded individual parcels of land to indigenous families, while later concessions were granted to legally organized and registered communities. Finally, in the case of the Huaorani nationality, indigenous lands were titled without the group having to be recognized legally (Uquillas 1993: 185). However, ancient boundaries between communities are being contested in practice by the colonizers. Encouraged by semi-directed colonization projects, poor peasants from Andean or coastal rural areas have entered the Amazon in relatively large numbers (E. Salazar 1981).[17]

Couched in terms of the rights inherent in citizenship, indigenous landclaims

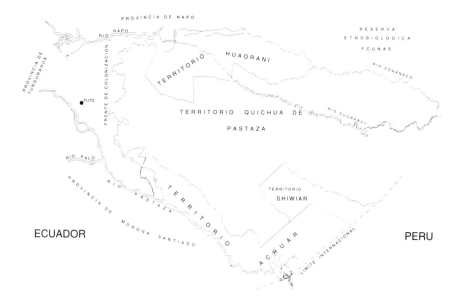

Figure 5.4 Map showing extent of landclaims under OPIP proposals from early 1990s

utilize the discourses of the state in order to challenge it. One topographer in Puyo explained, 'well, we as Ecuadoreans have full rights to lobby the national government, so that they give us a hand, support us'.[18] Nevertheless, the appropriation and deployment of cartographic techniques in the judicial system has not been without its ironies and contradictions. The two-dimensionality of the topographic maps bypassed and silenced issues of state control over the subsoil, claimed as part of its inalienable sovereignty. Official land titles do not give subsoil rights to indigenous nationalities; as elsewhere, legal title gives ownership of the land surface, not its depths. Indigenous communities thereby remain obliged to permit all types of mining and oil extraction activities on 'their' territories. In this respect, as one CONFENAIE leader acknowledges, 'the granting of titles is not the solution to the problem of indigenous territories' (Grefa 1993). Indigenous political activity has thus shifted to dealing with the multi-national oil companies and their local representatives (Kimberling with FCUNE 1993).

The topographic work carried out by indigenous confederations illustrates the malleability of cartography's use, and some of the politics in its usage. Cartographies are not tied immutably into colonizing and projects of control, but can offer ambiguous spaces for re-identification. Since the late 1980s, the mobilization of indigenous groups for landclaims reveals the shifting political agendas around cartography and the possibility for its subversive appropriation. Yet such appropriation (occurring within the context of the state legal system, the state's sovereignty over subsoil, and the increasing globalized economy of the region) is restricted by the meanings, powers and increasingly de-centred nature of control over land. Amazonian land surfaces have been nominally allocated to indigenous villages and nationalities. However, the very transferability of cartographic practice did not provide closure and permanence for indigenous geographies due to the ongoing issues of subsoil rights and oil extraction procedures. Resistance to state conceptualizations of national land and livelihoods engage contradictorily in debates around landownership and use, about which community is to benefit from 'development', and the social boundaries between commmunities. Confederations have attempted to interpellate indigenous identities within the discursive framework of a re-negotiated, more inclusive citizenship, while settlers and companies act on the basis of formal citizenship rights, and official norms. Indigenous strategies have emphasized the importance of territorial boundaries (marked on the ground and on maps) as the basis for community maintenance. Yet both projects interlace with identities and practices conjoined by the state and indigenous groups simultaneously, reconfiguring both hegemonic and subaltern national communities.

INTERNATIONAL AFFILIATIONS

Ecuadoreans perceive themselves in outsiders' eyes as well as in light of their own representations; the screen held up to Ecuadoreans by external images of the country provides a resource for the expression of ideas about identity and nation for many of our respondents. Official ideologies of nationhood, which stress the *internationality* of nationalism, are also drawn upon in popular senses of belonging, such as in opposition to Peru (see Chapter 3). There are a diversity of arenas through which a sense of uniqueness (although often based on stereotypes) can be expressed. The first dimension of this is that Ecuadoreans expect outsiders to highlight the natural features of their country. Natural features are perceived to represent the country externally; products (such as bananas, oil or shrimps) and landscapes are seen by popular subjects as elements through which foreigners identify their country. In Coca, one person said, 'It's a small country but rich in honesty, which the foreign companies recognize'. Moreover, official ideologies of nationalism focus in on certain features (the Galapagos, the mountains of Chimborazo and Cotopaxi for example) as natural sites unequivocally 'standing in' for the nation.

Second, Ecuador is positioned within a 'league table' of countries, on the basis of its economic development. Given the government's emphasis on the importance of homogeneous economic and 'social' development throughout its territory (a relation which is also gendered: see Chapter 6), Ecuadorean subjects perceive themselves within a 'ranking' of nations. Feeling embarrassed or angry about their country's low ranking in this international scale, Ecuadorean respondents self-consciously contextualized their identity in this framework. Many respondents felt that their country is perceived as a poor, *atrasado* ('backward') nation, characterized by its poverty and unequal income distribution. Others felt that this backwardness prevented Ecuador's true value being recognized elsewhere. 'Ecuador is not known because it's backward', said one *mestiza* woman in Saquisilí; the irony of this statement rests on Saquisilí market's growing importance as an (international and national) tourist attraction.

Given the official emphasis on Ecuador's Amazonian region, it is perhaps not surprising to note that among certain groups it is the country's Amazonian identity which is known internationally. One young *mestiza* female student from the town of Tena, Napo, said that Ecuador was represented in the world for 'being an Amazonian country'.

Other identities, which refer to supranational spheres, are expressed by respondents and thereby qualify both the differences set up through outsiders' perspectives, and through official discourse. Through the idioms of religion and shared colonial histories, the possibility of Latin American unity and a common identity are constituted and discursively created. Religion, especially Catholicism, provides a means through which to imagine a supranational community for Ecuadorean respondents. Among Protestants and Catholics in the villages and provincial towns of Cotopaxi province, as much as in Quito's elite, religion was named as the prime source of unity and common identity among Latin Americans. One-fifth in Cotopaxi (21 per cent) claimed that religion and its associated culture united Latin Americans in a supranational community, although over half (56 per cent) declined to, or could not, name a unifying factor. Similarities in colonial histories, and in Hispanic language use, were factors mentioned by a further 12 per cent. Oppositional identities were expressed by members of Quito's economic elite, who identified a fragile Latin American community in contradistinction to the 'other' of the United States, the International Monetary Fund (IMF) and the European Union. As one Quiteña white woman said, 'there is beginning to be a feeling that it's necessary to unify against the great powers such as the IMF, the European community, the United States'.[19] Among younger subjects, increased participation in international circuits of taste in music, clothing and lifestyles were perceived as more significant.

In relation to supranational identities, our respondents were more secure in their discussion of factors which are constantly relayed to them via government discourse (position in development, natural features) or in the longterm sites through which Latin American identity has been articulated (religion, language, and anti-US sentiment). Yet the majority of respondents saw only limited, or fragile, unity between Latin Americans, conditioned by their knowledge that the structures of government and the effect of divisive national discourses is to highlight differences rather than commonalities between the region's nations.

Rather than having identities bounded by the territorial borders of the modern nation-state, material from around Latin America suggests that identities are constituted in relation to multiple geographies of identity. The constellation of sites and socio-spatial relationships through which (national) identities are made draws attention to the contingency and tenuousness of affiliations with the 'national place', as if it was a pregiven boundary with which to identify. Collective subjects, in addition to individual expressions of subjectivity, are expressed contingently in relation to different sites and spaces, in which affiliation is expressed in the *juxtaposition* of places and their meanings. In the case discussed above, being Peruvian for second-generation migrants in Lima is as much about renegotiating discriminatory regional stereotypes and the value of popular, hybrid cultural forms, as it is about adopting unquestioningly any prevailing official versions of national affiliation. Affiliations with multiple places are then cross-cut by relations of class, location, gender, age and 'race', in which a sense of belonging is mediated through these power relations and positionings. Popular everyday expressions of relationships to place and national space are in themselves multiple, frequently contradictory and contested.

Moreover, a sense of national identification and belonging is always already conditioned by local and regional affiliations. In other words, local senses of belonging fundamentally condition or qualify the national identities which are created and re-produced. Localized expressions of national affiliation are often expressed through means of local/regional aesthetics, practice or materials (for example, the ingredients for Sicuani's red and white *trago*) which, through their disposition, origin or use, display the elements of local identities not replicated at the national level, or which are differently expressed at the national level. In other words, there are no unmediated national identities – any identifications with the national are mediated profoundly by local/regional affiliations (anything from neighbourhood, village, district to region, although less at the supranational level). It is in the particular configuration of the local *vis-à-vis* the national in discourses about rights, morals, aesthetics, economies, societies and so on which become the grounds for conditioning an affiliation with the national. Sicuani is happy to identify itself as Peruvian in colouring *trago* like the Peruvian flag. But the choice of materials to colour the drink is one which *conditions* the national identification – it qualifies it, and sets it within a discursive frame placing 'limits' on the national identity of Sicuaneños, at the same time as expressing positive identity and affiliation with a unique local place.

Multiple geographies of identities can be said to refer to the contested spatial dimensions of the multiple processes of identity formation and affiliation to place. In Latin America, geographies of identity reflect on – and build collective and individual subjects through – the interconnections between *patria chicas* and the global economy, between *barrio* neighbourhoods and regionalisms, between domestic/familial places and the official national monuments mapping out national places. A multiplicity of places (imagined through specific places and spatial relations) give meanings and nodes for identification to citizens in Latin American countries. Yet certain boundaries are still maintained and reified, despite this movement and malleability in geographical imaginations and affiliations. For example, despite the common ethnic heritage of groups on either side of the Peru–Ecuador Amazonian border, the majority of popular subjects – including those self-identifying as indigenous – identify with Ecuador and feel themselves to be Ecuadorean citizens, not questioning the affiliations that the international boundary demarcates. To give one example, a 24-year-old Catholic Quichua woman in the Oriente claimed not to know Ecuador, but correctly identified all the national symbols and said she was an Ecuadorean citizen 'because I was born here'. Another Quichua-speaking man of the same age, in Tena, rooted himself in relation to his birthplace and his ancestors, 'Yes I am part of the nation because I was born here, and for the inheritance left by our grandparents'. A 'mosaic' of places, and the communities associated with those places, are then the means through which identification and affiliation with nations can be expressed. Just as importantly they are the means by which claims to national belonging can be contested and renegotiated. Certain territorial boundaries are recognized as significant and as primary in self- and others' representation – for example, the international boundaries between countries. Yet the contextualization around these relatively 'fixed' social places – provided by regional, local, neighbourhood and supranational affiliations and the connotations of each of these – provide a route into the renegotiation of the power-filled and often officially sanctioned versions of identification with the nation. However, these identifications are of course importantly structured around gender and gendered identities. Discussion of gender and sexualities, and their impact on national identities, is taken up in Chapter 6.

6

GENDER AND NATIONAL IDENTITIES

Masculinities, femininities and power

In Latin America, nations and associated modernist discourses of democracy have been filtered through ideologies of gender (Luna 1993). Powerful discourses of nationhood link women and men to the nation in highly gendered ways, simultaneously fractured through complex relations of sexuality. Being one key effect of power, gendered bodies are subject to the creation of 'nationalized' behaviours and representations. Such relations shape negotiations over citizenship as women and men have distinct places and identities through which to express claims. Spiralling through the national imagined community, the masculinities and femininities defined as 'ours' are constituted opposition to 'Other' genders.

However, the multiplicity of masculinities and femininities existing in the national space provide no closure for official accounts of 'national' men and women, being constantly reconstituted and reformulated. Since the early 1980s in Latin America, changes in the 'imagined communities' and lived practices of participants in new social movements (especially in women's social movements and the emerging homosexual movement) have rearticulated national identities. The complex linkages between gender and nationalism have often been overlooked; as Anne McClintock (1993a) reminds us, theories of nationalism have tended to ignore gender as a category constitutive of nationalism itself. Looking at Latin American material, this chapter examines the ways in which sexualities and gender identities are constituted in the nation, and the ways in which national power intervenes to differentiate citizens on the basis of gender. In this approach we recognize similarities with what Anne McClintock calls a feminist theory of nationalism, which understands nationhood in terms of gender-aware theory, bringing into focus women's active participation in national formations, and paying careful attention to racial, ethnic and class structures which interact with gender (McClintock 1993a).

State gender ideologies are explored in the first section, while the formation of national gendered citizens (second section) is examined as mediated through 'race' (third section), bodies and embodiments (fourth section), the re-presentation of national memories and histories (fifth section) and national geographies of identity (sixth section). The rearticulation of gendered national identities is rearticulated in the continent's democratizing social movements (seventh section). The final section draws together the discussion by summarizing the chapter's main points.

STATE GENDER IDEOLOGIES

In what Joan Scott (1986) calls the gender ideologies of the state, nationalism is constituted, from its origins, as a highly gendered relationship, dependent upon the marking on women's and men's bodies and in their identities, of the ideologies of national difference. Women are not 'imagined', in Anderson's (1991) suggestive phrase, to be national citizens. The rise of bourgeois republicanism in Latin America was linked with 'limitations . . . for creating or imagining women as subjects of history' (Pratt 1990: 49), so nations have been explicitly linked to men and masculinities. Nationalisms – and generally national identities – are imagined *brotherhoods*, a 'horizontal comradeship' (Anderson 1991: 16) between men, who are simultaneously engendered as masculine and produced as national subjects, in the 'sanctioned institutionalization of gender difference' which is nationalism (McClintock 1993a: 61; also Kandiyoti 1991b). Nationalisms have been seen to have 'typically sprung from masculinized memory, masculinized humiliation and masculinized hope' (Enloe 1989: 44, quoted in McClintock 1993a). Nationalism acts to mobilize and interpellate groups for the governing project, and so mobilization of women and men as gendered subjects has been fundamental to nation-building, as the multi-faceted identities called up by the national project must include *gendered* subjects who can recognize themselves in the mirror of the imagined community.

During Latin American history, states' gender ideologies have informed policy decisions and the ways through which women and men are drawn into the state project of nation-building (also Alvarez 1990). Particularly in Latin American discourses around the 'modern woman', states have identified women in the role of reproducers, as social agents in community development, and domestic crisis-management agents (Luna 1993: 14). Depending on a country's political and social relations, the state mobilizes female subjects calling upon various identities and practices in its nationalism, as can be seen in Argentina, Mexico and Peru.

In Argentina, home of Juan and Eva Perón and the political ideology of populism, strongly gendered state ideologies of nationalisms have been negotiated over, determining both male and female citizenship roles. In populism's gendered language, *patria*, the nation, is symbolically linked to fatherhood whereby 'fathers'/leaders are 'consorts of the entire land, by extension' (Sommer 1991: 257). Virility is valorized as a male attribute (Sommer 1991: 266), prefiguring female passive subalternity in patriarchal national discourse. Juan and Eva Perón, in their efforts to mobilize gendered citizens, built on and refined these general patterns through specific practices and discourses. The Peróns wanted to attract women to their populist nationalism, and did so through granting women the vote in 1947, organizing the Partido Peronista Femenino (PPF: Peronist Feminine Party) in 1949, and imposing female candidates in the early elections (McGee Deutsch 1991). Yet the re-representation of female citizens in certain spheres relied upon the labelling of appropriate and unappropriate female characteristics. Being 'anti-nationalist', feminists were not role models; Eva Perón prefered to emphasize women's feminizing social action and women's subordination to their husbands (including her own). Yet Peronism's gender ideology was unpredictable in other aspects: male *peronistas* were encouraged to be humble, loyal and obedient, while Eva Perón exhibited strength and authoritarianism (McGee Deutsch 1991: 277). Overall, however, traditional gender relations were reinforced as Perón himself – the 'First Sportsman' of the nation – kept in shape through sport, and encouraged male sports. Certainly, male citizens appear to have interpellated such ideologies; 'With Perón, we were all *machos*', one labourer said (McGee Deutsch 1977: 279).

In Mexico, the state's gender ideologies were distinct, shaped by the presence of large indigenous populations, and the country's specific political trajectory. In early nationalism, indigenous femininity symbolized treachery and conquest of nationhood, with the rich imagery of La Malinche simultaneously representing gender, 'race' and colonial subalternity. By the time of the Mexican revolution, women generally were perceived by the revolutionary spokespeople to be the ones 'taming' rebellious and unruly men. Women thereby gained an (albeit passive) role in the revolution, then were associated with social conservatism which made male legislators – inheritors of the Revolution's ideologies – reluctant to grant women the vote. Consequently, women in Mexico gained suffrage rights at federal level only in 1955 (McGee Deutsch 1991).

In the Andes, where indigenous populations retained a relatively high degree of autonomy over household relations and village politics, the interaction of state citizenship norms with male communal hegemony resulted in a complex citizenship for indigenous Andean women. In Peru until 1969, suffrage rights were restricted to literate nationals, and due to high rates of illiteracy among indigenous groups (particularly women), Andean women had restricted access to 'public' political rights and formal representation. Discrimination by the *mestizo* state, in addition to highly complex gender relations *within* indigenous groups themselves, explain this pattern. In Peruvian Andean areas, women symbolize (through clothing, monolingualism, and bodies) a culture formed in resistance to the *mestizo* nation. Moreover, local patriarchies enforce female marginalization from the public sphere (through male domestic violence against women), and female disempowerment in household relations and educational access (Bronstein 1984; Radcliffe 1990a). The interaction of 'communal hegemony' (Mallon 1995) with gender-informed state politics together worked to reinforce local gendered citizenship, relations in turn saturated with 'race' and class. However, Peruvian indigenous women rearticulated their discursive positioning in the nation and in indigenous groups, thereby contesting communal and state hegemony. During the 1980s, indigenous peasant women organized through, and alongside, the Peruvian peasant confederations, demanding a greater say in decisions and a right to land and leadership. To symbolize their agenda, they utilized the figure of Micaela Bastídes, who had been a co-leader with her husband Tupac Amaru in an uprising against colonial authority in the 1780s. The figure was mediated by the corporatist military regime's re-presentation of Micaela Bastídes in the 1970s, which used this symbol to demonstrate a commitment to peasant rights and a shift away from creole dominance. In their campaigns, Andean peasant women reconverted the polyvocal symbol of Micaela Bastídes once again, to represent their demands for equality along-side their husbands in union structures. What had been an *indigenista* move by an urban military government became in the 1980s the hybrid means through which indigenous rural women attempted political authority and a decision-making voice (Radcliffe 1993a). Gendered national identities are, then, articulated not only in relation to state gender ideologies, but also in relation to the public ideas of citizenship,and other contradictory sites of identity production. Such gendered state ideologies raise issues of the contexts for men and women's citizenship and participation in decision-making.

GENDER AND CITIZENSHIP

Discourses and practices of citizenship have been deeply embedded in the constitution of modern nation-states, whose gendering of citizenship has recently received more scholarly attention. According to Enlightenment political philosophy – which has deep roots in Latin America – republican government represented the sphere of rationality and public state citizens. By contrast, the 'private' sphere of affection, spirituality and the family were separate (Pateman 1988; Marston 1990). Consequently citizenship was perceived as inherently masculine, based as it was on the dichotomies of male/female, public/private. The 'fraternal social contract', upon which the modern nation-state is arguably based (Pateman 1988), defines the nature of both public and private spheres as male-dominated. The term patriarchy is often used to describe male dominance of public and private spheres, although it takes different forms over time (Walby 1990). Historically, individual patriarchs controlled women and their labour in private domestic patriarchy, as occurred in (especially rural) colonial and early republican Latin America. For example, in Brazilian plantations, male heads of households exerted extensive personal and political control over their family and household workers, despite social change elsewhere in the country (Stolcke 1988). Under these circumstances, it was argued that the effect of public patriarchy was not so significant (Walby 1990). However, evidence from nineteenth-century Latin America suggests that male control over women was exerted through a variety of *public* means including institutionalized norms and practices, even in the 1840s. In early republican rural Nicaragua, court decisions and legal provisions upheld 'private' patriarchy, thereby breaking down boundaries between familial and 'public' male control (Dore 1996). Such work suggests that 'we should be wary of ethnocentric definitions of the private and the public' (Kandiyoti 1991a:430).[1]

In Latin America, the two gendered spheres of public and private are a central part of cultural resources used to explain and order social life, and through which relations with citizenship rights and duties are mediated. The public/private division is expressed through the *casa/calle* (house/street) distinction. The *calle* was the arena of formal politics, of the negotiations and alliances between men which defined, legislated and enforced rules of suffrage, participation in the labour force and international policy. The *casa* historically was a sphere with no legislated rights for female subjects; wives were often required to request their husbands' permission to work outside the home, under the *patria potesta* (laws giving power to male family members over women), on the statute books until the mid-1970s in many South American countries. Similarly, property was legally under the male household heads' control, while divorce was often restricted or illegal (although variations between countries occur, between Ecuador where divorce was permitted from early in the twentieth century, and Argentina where divorce was permitted only from the 1970s). Yet *casa*-informed legislative frames underplay the importance of 'national domesticity', through which national identities were apprehended and reproduced. As private citizens, women were to provide spiritual values and moral guidance to their children and to menfolk, granting stability and enhancement to the public sphere. In other words, women's power derived from their 'production of tradition' and morality, through which their citizenship is granted (Luna 1993: 10).

Critiques of the public/private divide draw attention to the ways in which the two spheres are interconnected. In rural areas where production and reproduction tasks are not easily separated, and in the increasingly globalized economy (with a resulting proliferation of domestic-based manufacturing), the dependence of one sphere upon patterns of domination, labour organization, and resource-allocation formed and reproduced in the other is clear (Wilson 1993; Women and Geography Study Group 1996). Domestic service persists as a major labour market for women in Latin America (Cubbitt 1995), through which relations of patriarchy, racism and nationalism interact to domesticate and replicate wider social patterns (Radcliffe 1990a; Chaney and García Castro 1989). Identity formation, occurring in intersubjective relations, thus takes place through relations whose boundaries and contents are constantly shifting and being negotiated. Moreover, *casa–calle* divisions depend upon certain class and 'race' specific gender relations. Through the early republican period, it was only elite creole men who had suffrage and formal citizenship rights; when female suffrage was introduced, it was often restricted by class. In the complexities of local cultures, the interaction between different classes and distinct patriarchies can result in gendered citizenship relations which cross-cut simple public–private divisions, such as in the Peruvian example given above. In the Caribbean due to highly diverse local societies, the *casa–calle* distinction had limited applicability in class and racial terms, as in many Latin American areas with diverse multi-diasporic populations.

RACE, GENDER AND NATION

When asked which ethnic group she belonged to, Miss Ecuador declared herself to be 'white', reproducing the perceived close links between beauty and whiteness held by many Ecuadoreans. The women who gain most prominence in the Ecuadorean press and television are highly racialized, being fair haired, blue-eyed and fair-skinned. Such public images of women refer specifically to the elite and upper-middle urban classes, reinforcing divisions among women on racial and class grounds, reinforced by media tips on makeup, 'Western'-style clothing and bodily form. Among Ecuadorean women, notions of beauty and self-esteem are constituted in relation to these racialized images of womanhood. Among women in the highland Andean community of Zumbagua, women brought up in indigenous culture say that they are ugly, 'just *runa*', and hence 'worthless' if they do not have 'white' facial and bodily features. *Suca* (beauty) is whiteness; the pairs *runa/suca*, ugly/beautiful, worthless/valuable are internalized from a young age (Weismantel 1988). The multiple discourses through which gender difference is apprehended and reproduced means that ideas about race/beauty reproduce gender *and* national identities, by specifying which feminine criteria are the focus of the imagined national community. The everyday presentations of valued female racialized beauty reinforce ideas of who is the nation, and who is valued/ desired by the nation.

Xuxa, the Brazilian television star on Brazil's largest channel, has, since the 1980s, re-presented to the Brazilian people their own tensions and thoughts about gender and 'race'. Xuxa's blonde beauty reinforced one particular femininity as desirable in Brazil, despite her very uncommon appearance. With her constant body-grooming, Xuxa represented a first-world oriented consumerism, available only to the elite yet aspired to by many women. Women measured their appearance against her as their ideal; 'I may not quite be Xuxa, but I look better', said one woman after liposuction work (Simpson 1993).[2] In a country with the second-largest number of black people, Xuxa's blondeness and north-European features would seem to be totally anomalous, yet by buying into the national myth of racial democracy, Xuxa's racial characteristics placed her on a continuum ' that informs Brazil's cultural geography, within which the country is divided into a predominantly white south and a nonwhite north' (Simpson 1993: 15). In Xuxa's case, the ideal- ization of femininities occurs under the self-conscious gaze of the nation. Brazilian self-representations refer more to the ideal of whiteness, than to the *mulattas* and black samba dancers relayed to foreign tourists and outsiders (Simpson 1993: 39). In Belize, ethnic diversity is celebrated self-consciously, placing more emphasis on the female body 'stripped' of ethnic affiliation and saturated with nationality (Wilk 1995). In the case of beauty queens and stars like Xuxa, 'race' crucially intervenes in the discussions of gendered roles and reproductive sexuality (see also Westwood and Radcliffe 1993).

In eugenics debates circulating in Latin America from the end of the nineteenth century until the 1940s, questions of male and female reproductive roles came under much scrutiny, particularly since nations were increasingly imagined in racial terms. As Leys Stepan (1991) points out, eugenics vaunted 'racial purity' during the interwar years thereby creating 'explanations' for rapid socio-economic change. Reproduction and health were pinpointed as key policy areas, and measures were debated which could (it was argued) influence the national community's makeup through controlling who reproduced, especially by focusing on women's reproductive capacity. In Latin America, 'negative eugenics [that is, the control of marriages] made women's fertility seem a crucial resource of the nation, thus locking women into reproductive roles' (Stepan 1991: 122). Especially in Argentina (where the 1930s were characterized by racism and fear of a 'crisis'), various proposals were made for controlling marriages, proposals whose main object of control was 'woman' (often marked by class and 'race') but also men and their sexual urges. Proposals included creating a register for easier management of the national 'biological patrimony' (Stepan 1991: 119), sex education to 'subdue the sexual instinct', and control of homosexuality and sexual licence (Stepan 1991:118). Female reproductive capacity was of particular concern; Argentine eugenists proposed

> an involuntary state system of surveillance of motherhood and the forcible registration of pregnancy, since pregnancies represented to them the vital germ plasm of the nation.
>
> (Stepan 1991: 121)

The Argentine case is particularly extreme, and these proposals were not implemented. Yet Leys Stepan argues that these notions entered the public domain, influencing maternal health policies and marriage law. In several Latin American countries, couples had either to present a certificate to declare themselves fit to reproduce, or were refused marriage licences if they had disease or 'vices' that endangered future offspring. These rules applied in Mexico, Brazil, Peru and Argentina at various times in the early twentieth century.

Men, especially lower-class men, were viewed as particularly irresponsible and diseased; the Argentine prenuptial law applied only to men. Yet connotations of masculine uncontrolled sexuality are highly unstable and subject to reinterpretations which can make them either positive or negative national attributes. National identity expressed in terms of action, vitality, vigour and so on are most commonly associated with masculinities, male bodies and masculine sexuality. In other words, it is not only the physicality of female bodies (and their procreative aspect) which are appropriated and transformed by nationalist discourse, but also male physicality. Cenotaphs, tombs and plaques map out male heroism, providing an opportunity for showing national masculinities (and at times femininities) (Anderson 1991; Pratt 1990).

EMBODYING THE NATION

National masculinities and femininities are not 'merely' symbolic and discursive, as they engage with – having both productive and limiting effects on – bodies and the subjectivities embodied in citizens. Control of bodies – especially women's bodies – is central to the practices and discourses of nationalism. Yuval-Davis and Anthias (1989) point to the specific links between women and nations, listing five roles for women – as biological reproducers of national citizens; as biological 'markers' of boundaries between national groups; as subjects involved in the cultural transmission of national values; as signifiers and symbols of national differences; and finally, as participants in national struggles. Despite a problematic emphasis on racially defined nation-states, Yuval-Davis and Anthias' analysis emphasizes the embodiment of national agendas in engendered bodies, as well as the practices through which engendered bodies engage with the nation.

Moulded by national practice and discourse, women and men become embodiments of 'the national'. Behaviours, senses of self and identities, all are imagined and expressed through the body, which in Latin American nations is represented and lived as inherently gendered/sexed (Moore 1994). One way into analysing the embodiment, and associated re-presentation, of male and female citizens in Latin America has been the model of *machismo* and *marianismo* (cult of female moral superiority). Notions of *machismo*/*marianismo* revolve around the idea that family social relations in a Catholic society – informed by notions of honour and shame – form gender relations based on openly heterosexual and aggressive male behaviour, and on meek and self-abnegating female behaviour and chaste female bodies (Stevens 1973; Cubbitt 1995). Drawing upon the Mediterranean and Catholic colonial heritage, *machismo* and *marianismo* in effect engage with regional, gender or class identities rather than national difference at these levels of generality. In other words, everyday understandings and expressions of *machista* and Marian behaviour are as much to do with gender difference, class or race identities, and only tangentially to do with nationalism. While many grassroots women in Peru, Guatemala or Uruguay express their agenda in terms of dealing with *machismo* (Kuppers 1994; Radcliffe 1990a), only for some women does a Marian identity hold much reward. For example in Ecuador, 'while spiritualism, sacrifice and so on are perhaps characteristic of some groups of women (especially upper-middle class women), it is not true of all' (Stolen 1987: 57–9).

Yet the strongly maternal component of Marian identities is one frequently recirculated in Latin America, as shown in the Argentine and Mexican examples discussed above. Pratt suggests that in Latin American bourgeois republicanism, women were offered 'republican motherhood', a role in which their reproductive capacities would be respected within conventions informed by patriarchal, class- and race-defined honour and shame discourses (Pratt 1992). As national mothers however, they were not part of a 'national' male brotherhood but 'precariously other to the nation'. In other words, the Marian identity is shaped and rearticulated to become a symbol of *national* femininity (for example, figures of the Virgin and famous national women), where the always already ambivalent (non)national status of women reveals their marginality within the national project. Mary, the Virgin Mother Catholic symbol, is perhaps the most ubiquitous and easily recognizable female emblem in Latin America. Drawing on centuries of representation in Latin America and Europe, the Virgin Mary has become deeply embedded in numerous local cultures of the continent, from elite drawing-rooms to peasant chapels. Despite its ubiquity and universal meanings, Virgin figures have also – in their polyvalency – come to express nationalisms, through the specificities of local usages and characteristics attached to this relatively 'ungrounded' symbol. The histories invented for 'local' Virgins create points for local identification at the same time reproducing features of (universal, desirable, ideal) womanhood. The Virgin of Guadalupe, patron saint of Mexico, illustrates the double instability of local/global Woman symbols for nationalism. First appearing to an indian in 1531 on the site of a pre-Colombian shrine to the 'mother deity' Tonantzin, Guadalupe was soon worshipped by creoles and *mestizos*, and through a combination of its local (largely indigenous) and national (mostly creole and *mestizo*) meanings, became an emblem of identity for the 'people', rallying Mexicans to fight Spain in 1810 (Rowe and Schelling 1991). A century later in the Mexican Revolution, Zapatista rebels also used the Virgin of Guadalupe as a unifying icon against a discredited state.

In familial allegories of nations, secular women can also be appropriated by national narratives, as in the case of the 'Mother of Cuba', Mariana Grajales Coello. Although a recognized historical figure in the nineteenth century for backing her sons' anti-Spanish struggles, Mariana Grajales was elevated to a national symbol only in 1957 when she was officially declared 'Mother of Cuba', 'a symbol of abnegation and patriotism' (Stubbs 1995). Confirmed through statues, a national day and eulogistic studies, Mariana Grajales came to signify maternal concern, Catholic religiosity and patriotic fervour, in a country where *marianismo* is a strong ideology but where Catholicism has been replaced by communism as the official guiding ideology. Although herself a free *mulatta*, Mariana Grajales was utilized to symbolize neither black liberation nor the value of *mestizaje*. According to some interpretations, Mariana's own marginality (free coloured, Dominican-born) may have made her a more malleable symbol for the nation (N. Redclift, personal communication, 1994). Certainly as a sacrificing mother icon, *marianismo*'s regionally free-floating sign became grounded and 'nationalized' through association with a *mulatta* who had been actively involved in national liberation struggles. However, such an interpretation underplays the ambiguity and contradictions in this female nationalist icon, which revolve around the fact that a real historical female subject *does* appear in nationalist historical narrative (as do women in the Ecuadorean press, yet that real female subject is 'made over' in the national narrative. This 'making over' highlights only one dimension of a multi-faceted identity – that of gender – whose stereotyped attributes (sacrifice, maternal love) saturate other dimensions of subjectivity. With women 'imagined as dependent rather than sovereign' (Pratt 1992: 51), women throughout Latin America have been granted a position at the margins of nationhood, in which their sexed bodies (gendered through ideologies of *marianismo*) saturate the multiple other facets of their identifications with place and nation.

Yet calling upon gendered and sexed national bodies is very variable, differing with the state gender ideologies in which they are embedded and prevailing ideas about a desirable national community. These themes can be illustrated by looking at the representation of masculine embodiments in Ecuadorean nationalist discourse; and modern Argentine masculine leisure activities.

Embodying heroism in Ecuador

An example is the plaque placed outside the *Casa de Sucre*, General Sucre's house in Quito, Ecuador. Erected in 1979, the plaque illustrates themes of masculine embodiments, male homosociality and national affiliations. The plaque reads,

> In the place dedicated to the immortal memory of General Antonio José de Sucre, at the foot of Pichincha [mountain], where valour, heroism and virility, particular virtues of Spanish and Ecuadorean [men], gave birth to a young nation, sister of Spain, HOMAGE is given to the SPANISH Soldier, paradigm of virtues of the Race, discoverer of worlds, defender of the [Catholic] faith, founder of cities, explorer of lands, creator with his blood of the new peoples of America.
>
> (Author's translation, capitals in original)

Beyond the by-now widely known and recorded theme of male soldiers' martyrdom,[3] the Ecuadorean plaque reminds us that this figure has a particular - contextualized embodiment in national gender relations. In the Sucre plaque, reference is made not only to General Sucre as independence hero, but also to Spanish conquistadors. What unites these men are their physicality and emphasized masculinity as 'originary' heroes. In the plaque, men (conquistadors and independence fighters) are represented as giving birth to a young feminized nation, a nation subject to male dominance arising from precedence and patriarchal gender relations. The nation's 'birth' takes place without women appearing at all, although they are the hidden sign in the creation of 'new peoples' (*pueblos*) mentioned at the end. Despite the wars of independence which pitted colonizer against colonized, Spanish and Ecuadorean men are represented as brothers with identical heroism, valour and virility, an example of a 'horizontal brotherhood' extending even *across* national borders in this case. Differences in nationality – which in other situations create overtones of gendered hierarchies of domination and subordination (cf. Silverblatt 1988) – appear in effect as superficial distinctions between men linked by 'domestic genealogies' (McClintock 1993a: 63) traced through their 'sister' nations and, more implicitly, shared norms of heterosexuality and homosociality. The creation of masculine heroes in Ecuador draws upon the unspecified (class, race) origin of men, just as in Costa Rica the hero Juan Santamaría was an 'anonymous' hero with obscure roots (Palmer 1993). Although not prominent, there are also implicit national roles for women in these monuments. The Ecuadorean plaque implies that women are to reproduce citizens (the 'new peoples'), while Costa Rican nationalism generated a 'satellite role model for women', in which women were encouraged to raise patriotic sons like Santamaría, training them in civilian duty at a tender age (Palmer 1993: 68).

Argentine leisure activities and national bodies

The ways in which certain embodied dispositions and sex-gendered bodies are associated with national identity can evolve over time, in relation to certain desired characteristics. In early-twentieth-century Argentina, masculine national embodiments were bound up with the rearticulation of *gaucho* masculinities. National attributes of physicality, sexuality and brotherhood all come to the fore in male leisure actitivies in early-twentieth-century Argentina, where tango, football and polo referred in subtly differing ways to the historic masculine model/body of the *gaucho* (cattle worker in Argentine plains). During this self-consciously modern period, football and tango – each with their own sites and social relations – were providing new bodily experiences and points of identity for a rapidly growing migrant population, thereby rearticulating their adopted nation and their place in it. In tango, new gender relations were negotiated and expressed, as men experienced new sensualized relations and women left domestic spaces for the cabaret (Archetti 1994b). Expressing ideas of male unrequited love for a self-consciously elegant woman, tango lyrics are interpreted by Archetti as a male call for a 'return to the forms of love and chastity associated with traditional family roles' (1994b: 112). Yet however traditional the subtext for tango lyrics, the new bodily experiences of the tango offered forms of identification for newly arrived migrants as an 'Argeninian' practice. (Official measures included obligatory conscription for immigrants' sons and the creation of nationalist school curricula, to instil nationalism among the new populations: Archetti 1994c; Rouquié 1987.) As such, tango – and in it the male *compradito* figure – represented a more open and flexible node of identification than the *gaucho*, whose origins in the *pampas* (plains), creole origins (Spanish-indigenous mixture), hostility to the state, and attributes of romanticism, generosity and boastfulness all made it an unlikely masculine model for generally low-income, urban, non-creole and upwardly mobile immigrants.[4]

Polo re-converted and hybridized masculinities and masculine embodiments in distinct ways, by 'civilizing' rugged and violent *gaucho* equestrian games and, equally significantly, making Spanish (not English) the game's working language. Between Argentina's first polo match in 1876 and the 1920s, polo was 'nationalized' after early British hegemony (Archetti 1994c). Yet *guachesque* characteristics persisted, providing a robust and physical masculinity through which national pride could be reasserted, especially in opposition to other nationals. In 1924, a player explained Argentine players' superiority in terms of their horses, resistance, and because

> we have much more temperament than [English or American players], we are real fighters due, perhaps, to an atavistic gaucho instinct which is still in our veins and is manifested in many of the things we do.
>
> (*El Gráfico* 1924, quoted in Archetti 1994c: 18)

However, it was mostly through football that a hybrid masculinity was created, offering sites for identification with the nation regardless of origin (see Chapter 4).

> The native Argentinians and the immigrants accepted and incorporated football as an important bodily practice in the use of leisure time and as a ritual context for competition and emotional display of loyalty and engagement.
>
> (Archetti 1994c: 13)

Regardless of class or ethnic origin, men could become embodiments of national values and traits and could *be seen* through increasing press coverage. Such a hybrid masculine embodiment was expressed in a playing style which was 'not Italian or Spanish, [but] "creole", because [it] developed from sources of mixed origins' (Archetti 1994d: 15).

Argentine masculinities in sport were then modern creole *gauchos*, savages only when required. Polo, football and tango were the bodily practices which, although carried out by highly specific and markedly sexed/gendered bodies, were embedded within wider social notions about bodies *per se* and the appropriate practices to display national belonging, sexuality and pride. Such representations resonated with the modernist discourse about bodies found throughout Latin America at the turn of the century. Modernist biopolitics was expressed through increased press coverage on physical education for women and men, notions of hygiene and diet, and the importance of sporting hobbies, even for the president (Archetti 1994d). Love of football and tango was not only for men in Argentina; women also could express through body language their adoration of sporting and dancing stars, as for example in receptions for football teams at Buenos Aires harbour, or on the dance floor. To summarize, the specific contextualized relationships between male and female embodiments are a key site for the expression and reproduction of specific national identities.

SEXUALITY, GENDER AND NATIONHOOD

The interaction between male and female in nationhood is often grounded on assumptions of heterosexuality, and nationally beneficial reproductive behaviours. Broadly speaking, women function as dominated reproducers of national citizens while male sexuality is channelled into reproductive forms, and homosexuality is curtailed (A. Parker *et al.* 1992: 6–7). Following from feminist theories about heterosexuality and patriarchy (Connell 1987; Foord and Gregson 1986), sexualities can be said to inform gendered power relations and national identities simultaneously. Theoretically, 'idealization of motherhood by the virile fraternity would seem to entail the exclusion of all nonreproductively-oriented sexualities from the discourse of the nation' (A. Parker *et al.* 1992: 7). Yet the presence of male bonding in national identities focuses attention on a complex multiplicity of male–male relationships in nationalism. Following Mosse's original analysis which focused on national identities' 'distinctly homosocial form of male bonding' (Mosse 1975), other writers have examined the control and definitions of (appropriate) sexuality and emotional relations in national self-definitions. The denial and control of homosexual practices, for example, exists simultaneously with an emphasis on heterosexualities, embodied in hegemonic masculinities and emphasized femininities (Connell 1987). These sexualities and mothering images are grounded and reproduced through defining the specifics of national difference, with nation-states highlighting the self-conscious distinctions drawn between *their* mothers, sons, heterosexual relations and so on, and those of neighbouring nations. Such sex-gendered relationships are constituted through class and 'race', as illustrated by recent work (e.g. Stepan 1991) on the racialized nature of 'reproductive nationalisms', in which female and male national roles depend crucially on their 'race' or ethnicity as well as sexuality.

In Ecuador, as in other Latin American countries, public displays of sexuality are confined to heterosexual modes and oriented to presumed heterosexual audiences. In the daily and weekly press and on television, women are represented as objects of male heterosexual desire, directing their sexuality towards a male gaze and to reproductive functions. Broadsheet newspapers frequently devote sections to 'women and the family', reinforcing women's role as reproducers. Press reports highlight the positive attributes expected of women as reproducers, and the joy and sacredness of motherhood, while professional women are represented as finding motherhood the most satisfying part of their lives (*Universo* 10 May 1992). The domesticating discourses around nationalized femininities are imagined through complex discursive reference to the entire national territory (McClintock 1993a; Yuval-Davis and Anthias 1989). Women's sexuality is mediated through and re-represented through their role as reproducers of (patriotic) future citizens. A message to Ecuadorean women in a press article illustrates the complex 'domestic' space assumed in national agendas. An Ecuadorean woman, it was said, is 'companion in the home, wife, mother, worker, teacher and so on, [who] drives national development and contributes with new citizens for the *patria* at the same time as she fights for her personal realization' (*Universo* 8 March 1993). Numerous press and radio features emphasize women's roles as educators and moral guardians to the young in their care. Since 1970 the National Women's Directorate organized mother–child programmes, treating women in their reproductive roles (CEIMME 1994). That women recognize their limited domestic power base via maternal roles is highlighted by many women. One middle-aged *mestiza* woman in the Oriente described her personal power as limited to 'carrying the reins of the house'.[5]

Such domesticating representations link to the well-being and future of the national territory; women are tied to the nation's future as producers of future citizens; in this sense, women are very strongly tied to the future of the nation, albeit indirectly (cf. McClintock 1993a). Again such discourses differentiate between those femininities (often creole and urban) valued in hegemonic notions of gendered nation-building, and 'other' femininities. Discourses imply that not all women are perceived as equally valued mothers, differentiating between rural (read largely indigenous) and urban women, inferring that the former are having too many children relative to the latter (*Universo* 9 March 1994).

One social group often excluded from the imagined national community are homosexuals, although in practice attitudes and legislation concerning same-sex relations vary around the continent (Green 1991: 144). While Brazil and Costa Rica have relatively large and open displays of homosexual relationships, in Ecuador the gay movement is repressed both by violence and by cultural attitudes.[6] The Ecuadorean legal code makes homosexuality a crime, punishable with sentences over four years. While the law is not strictly adhered to, it justifies intimidation of those assumed to be gay, as well as violence and extortion by police and health workers. A clandestine gay and lesbian publication, *En Directo*, calls for legislative reform and changes in social attitudes, as well as increased collective mobilization among homosexuals. Situating themselves firmly within the nation, gay activists claim that 'the state . . . is obliged to oversee the wellbeing of the men and women which comprise it, . . . without distinction of sex, race, religious creed or inferences about private lives' (*En Directo* (1992) 5 (April): 3). Popular culture maintains an ambivalent attitude towards homosexuality, permitting the appearance of two gay men on a popular television talk show, while the HCJB Protestant radio station offers to 'cure' homosexuals.

In Latin America generally, only recently have feminist movements acknowledged and allowed space for lesbians (and gays), reflecting general social conservatism and the lack of grassroots links between the two movements (Saporta Sternbach *et al.* 1992; Vargas 1991). A growing gay and lesbian movement in Brazil from the 1970s illustrates the complex interaction of public discourse and space for organization. A relatively liberal Penal Code (with no mention of homosexuality) did not prevent widespread social discrimination and misapprehensions in Brazil. Homosexual subjects campaigned (successfully) against the health service's definition of homosexuality as a 'mental deviation', and for recognition of homosexuality as legitimate sexual expression (MacRae 1992). In addition, the gay and lesbian movements brought 'private' issues of sexual preference into the public arena as legitimate matters for debate and mobilization. Alongside such political activism, the 'carnivalization' of gender roles in the movement – as in Brazilian society generally – highlighted the arbitrary nature of traditional gender identities. Yet such internalizations and behaviours draw upon and re-represent *past* behaviours and memories, as a screen and mirror to contemporary practice, as in the example of Argentina's *gauchos*. It is to the nationalist utilization of the past and its gendered subjects that we now turn.

GENDERED MEMORIES AND NATIONAL TEMPORALITIES

Memory and history in national narratives are gendered, drawing upon and re-representing to citizens in various self-conscious ways masculine and feminine role models. As expressions of national imaginings and myths as well as fact, histories of nationhood are markedly gendered in their templates, often highlighting hegemonic masculinities and emphasized femininities – the national heroes and heroines. Men appear in the histories of battles, governments and as monarchs, whereas women appear as icons of national domesticity, morals and 'private' sociality. In this context, female and male nationalist iconography is widespread in Latin America in literary, textual, sculptural and visual forms (J. Scott 1986; McGee Deutsch 1991). Yet such identities are ambivalent, drawing upon figures whose multi-faceted differences to current gendered subjects generate contradictions.

The ambivalent temporality of nations engages *gendered* subjects, revealing interesting dimensions to the gendering of national narratives (cf. Kandiyoti

1991a; Verdery 1993; McClintock 1993a). By emphasizing the force of discourse in positioning women and men so differently in national projects and modernity, these arguments provide some interesting avenues to be explored in Latin America, where the idea of a strong trajectory into modernity is deeply rooted in national narratives (see also Radcliffe 1996b). As the media are engaged in national self-representation and popularizing national history,[7] recent Ecuadorean government planning documents, census data, and the national press from the 1980s through the 1990s can reveal these gendered memories and imaginaries.[8] The insertion of female historical figures into national narratives (especially in present-day press and government reports) reveals that, in Ecuador at least, re-representing female subjects is a complex process. On the one hand, women are called to interpellate modern national discourse and practice, yet on the other hand the roles through which women seemingly contribute to that nation are limited.

To record and celebrate past times in the republic, female figures appear in Ecuadorean press articles focusing on 'key moments' in national, largely post-colonial, history. Although coverage occurs throughout the year, many articles explicitly focused on Ecuadorean women around the date commemorating International Women's Day (8 March); however, these pieces tend to emphasize the Ecuadorean nation and women identifying with it, rather than the international dimension. Moreover, reports on historical women pinpoint women whose national significance is positioned *vis-à-vis* male heros. One piece (*Telegrafo* 3 March 1992) mentions named women remembered for their contributions to independence or republican development.[9] Especially in independence struggles, the women are positioned by their ties (via marriage, maternity or kinship) with men whose names are assumed to be immediately recognizable to readers. Manuela Garacoya was 'wife, mother and sister of heroes', as well as being an active nationalist herself and friend of the Liberator Simón Bolívar. Manuela Espejo was sister and wife to Independence heroes, also working in a Quito library and nationalist newspaper. Female historical figures in Ecuadorean national narratives are thereby linked symbolically with the domestic space, the socio-spatial relations of elite and middle-class bourgeois homes where political ideas were discussed in the salons and new print capitalism. However, there is also a wider implicit spatiality to these representations, which links these women with particular locations within the nation, specifically the largest cities of Quito and Guayaquil, both of which lay claim to being urbane, civilized and politically central places (Chapter 3). Lest this recognition of women as major actors in national independence be too prominent (for example, readers are reminded that Rosa Zárate's head was displayed alongside her husband's in the public square), press reports discursively construct heroines whose work for the *patria* did not threaten their femininity. The wars of Independence, argued one article (*Comercio* 12 March 1994) demonstrated how women could work 'shoulder to shoulder with men without losing their femininity nor putting their homes in jeopardy'. Another article, quoting a letter from populist leader and several times President Velasco Ibarra, suggests that 'real feminism' was to intensify femininity (*Comercio* 8 March 1992).

One of Ecuador's most popular heroines is Manuelita Saenz, wife of the Independence leader and, to the majority of Ecuadoreans surveyed, the most famous Ecuadorean woman. As helpmate and companion to Bolívar, Saenz represents a domestically based feminine ideal, also active in the 'birth' of the nation. A recently opened private museum, the Museo Manuela Saenz, builds on this popularity (which is not reflected in school books or 'formal' histories), the goal being to 'rescue the true image of our heroine Manuela Saenz'. The museum aims to 'research, rescue, emphasize and publicize the historic and cultural values of our country, . . . especially from the Independence period' (Museo n.d.), and presents Saenz's personal objects, letters, and paintings of her as a white/creole woman.

The 1895 Liberal Revolution is also remembered as a time when certain women make their appearance in the national narrative. Ecuadorean women, placed firmly within certain class and racial groups, are, for example, inscribed within a 'brilliant history (*trayectoria*) of struggle against exploitation and injustice' (*Universo* 6 March 1993). The article argues that the irregular forces of Liberalism pressured for the deletion of the term 'man' from the 1897 constitution. Such texts reproduce Ecuador's self-representation as a modernizing leader, the origin and initiator of 'modern' nation-state practices. The article reminds readers that Ecuador was the first Latin American country to grant women the right to vote in 1922 (usually dated to 1929). Ecuador's (patriarchal) nation self-consciously presents itself as a modernizing state, initiating and carrying through modern measures, such as the granting of female suffrage. While there is some truth to this, in comparison with Argentina for example, women are not positioned as the agents behind these political changes. Rather, as part of its (inevitable) modernization, Ecuador claims to be the precursor of Independence struggles in 1810 and the enlightened guarantor of female suffrage. Male-dominated discourse among journalists and media workers thereby reproduces male-biased theories of nationhood, at the same time reinforcing notions of gender difference.

Such gendered historical narratives have several implications for *contemporary* gendered national identities, by writing the national in particular engendered ways, highlighting women-linked-to-male-national-heroes as antecedents and models for modern women. Readers are reminded how 'these [historic] women constitute a vigorous example for the new generations' (*Telegrafo* 8 March 1992; also *Universo* 8 March 1993). In the same article, a workers' spokesperson reflected on the 'examples left by Manuelita Saenz, Ana Villamil, Rita Lecumberry, which serve as guides [to women today]'. The female figures are appropriated (in ways which highlight their individuality and active presence in history, thereby disarming feminist critiques), yet simultaneously transform them into the femininities most valued by the contemporary 'nation' (or its largely male popularizers). Moreover, only certain racialized women are inserted into history; white creole women from the major cities and most influential families are offered today, through newspaper reports, as exemplars to contemporary women. 'Nationalizing' of women – and gendering of national history – is thus an ambiguous process, drawing women into the modernity of the nation on specific (racialized and class-specific) terms. The way in which these racialized/classed women are identified and pinpointed also works through a complex gendered geography of identity, also identifiable through media and government discourse.

GENDER AND NATIONAL GEOGRAPHIES OF IDENTITY

Speaking about how economies grow and living standards improve if a 'nation's women' enjoy a favourable status, a *Universo* article argues that if, by contrast, women live in 'unfavourable' economic, educational and health conditions, development will be slowed (*Universo* 9 March 1994). Modernist discourse tends to see close ties between a nation's development and 'its' women's status, especially in countries countering Orientalist images (see e.g. Kandiyoti 1991b). In Latin America, such modernist discourses are re-encountered; in late-nineteenth-century Costa Rica, the Liberals linked the achievement of progress and development of science, art and education with ideas about women's emancipation (Palmer 1993: 67).

Ecuadorean discourses of development are expressed through quite different gendered figures to the national past. Generally, development is imagined as resulting in the homogeneous spread of capitalism and its modern features throughout the national space (see Chapters 3 and 5), while press and govern-ment documents make it clear that these imaginative geographies of national modernity are highly gendered discourses (constituted through 'race' and class). Ideas about progress are saturated with notions of the places and positions occupied by feminine (and masculine) subjects in the nation. We can suggest that the modern national project's contradictory positioning of women is an ambivalence constituted around place (mediated through class and 'race') (cf. Rose 1993). Although all women can be represented as contributing to development in similar ways, differences between women – in activities, 'race' and location – are also pinpointed, drawing certain women into the nationalist project more than others. The nationalist interpellation of women is reinforced through the geographies of identity infusing newspaper articles and government discourse (see Radcliffe 1996b).

Reflecting widespread belief in the possibility of women's integration into modernization (Kabeer 1994; C. V. Scott 1995), women's gender-specific position within Ecuadorean development plans can be traced back to the early 1960s, when debates about women in national development began. At this time, women's groups were set up under the guidance of social workers to instruct women in sewing, cooking, nutrition and basic hygiene issues (Andean Mission of the United Nations 1960: 13). However, it was in the 1970s – in the context of widespread national and international projects to integrate women more fully into the development process (Kabeer 1994) – that the Ecuadorean state's interest in women and development burgeoned.[10] At this time, working-class and peasant women in particular were the target of gendered welfare policies, such as the distribution of food staples (cf. Moser 1993b). Such programmes revealed an ambivalence about women's roles in national development: women were to be passive welfare recipients and maternal providers for families, and also more active participants in activities for modern development. Regarding the latter, skills training programmes aimed to train women in activities that can be seen as the epitome of feminine modernity – skills like hairdressing, beauty care, dressmaking, secretarial and accounting skills. Again presenting women as passive followers of the nation's trajectory, female integration into national modernity was founded upon a profoundly gendered division of activities – women's role in national progress was not to conflict with their domestic feminine role.

Women are interpellated through gendered geographies of identity, seen as complementary – and not threatening – to masculine geographies and activities. Participating in development is compatible with being men's home companion; 'Women's interest in contributing to the development of peoples [*pueblos*] and nations. . . . [is possible] without leaving off being the companion of men in the home' (*Telegrafo* 8 March 1994). The space for female citizens as development participants is thus posited to be the domestic place, providing reassuring continuity. This in turn had strong implications for the representation of rural development patterns.

Already in the 1960s, indigenous rural women were identified by the state as representing the potential limits to the development process. An international commission on the 'Indians of Ecuador' of 1960 reported that 'women in rural areas tend to a greater degree of conservatism and the retention of local customs, dress and manners' (Andean Mission of the United Nations 1960). To promote the integration of rural women into the development process, the Ecuadorean state instituted a number of rural development projects, drawing upon international notions of improving efficiency and human resources. Policies in the 1980s, such as those implemented under the FODERUMA programme, aimed to provide credit and technical assistance to rural women who were perceived to need technical skills training in sanitation, housing and nutrition (CEIS 1988). Similarly, the SEDRI programme in the 1980s involved women in small-scale income-generating projects and marketing ventures, especially in the Sierra region (in the provinces of Chimborazo and Cotopaxi). In the 1990s similar projects continued, such as those implemented under the FUNDAGRO (Fundación para el desarollo agropecuario – Foundation for Agricultural Development) project with national and Canadian support for twenty peasant women's small business enterprises (*microempresas*) to produce guinea pigs and chickens for sale, based in Cañar, Guamote and Colta in the Sierra (*Comercio*, 23 February 1994). Contemporary press reports echo this concern, drawing upon images and texts about poorer women, especially in rural areas.

While the international coalitions of governments, aid agencies and development organizations provide one context for these gendered programmes, the *national* narratives of gender, place and history also shape women's identities in development discourse. In government reports and associated press coverage, distinctions are made between femininities favourable for national development and those representing a potential obstacle to national progress. Differentiated on the basis of practice, place and 'race', indigenous women in the Sierra are discursively constructed as possible blocks to national development. As in the 1960s, rural women's positioning in national development narratives is based on the notion of a distinctive rural indigenous femininity. It is in relation to this gender/location/class/'race' specific group that the 'distinctive cultural topography' (Said 1994) of national development plans has been mapped. In comparison with other Latin American countries, (indigenous) rural women are not represented in Ecuador as the basis of national distinctiveness and a source of national pride. In contrast to Guatemala for example – where female indigenous clothing is seen to epitomize national identity, and is used widely in beauty contests and tourist adverts (Hendrikson 1991) – the Ecuadorean press and society value 'whiteness' and certain key activities in women (such as the urban 'pink collar' jobs mentioned on p. 151).[11]

In May 1992, press reports appeared in several newspapers identifying the specific situation of rural, especially Sierran women. Peasant women, 'although excluded by the Agrarian Reform', were seen as having to do agricultural work, especially in the Central Andes where 80 per cent of small farms were reported to be the responsibility of *campesinas* (peasant women) (*Hoy* 10 May 1992).[12] Other reports highlighted the higher rural rates of illiteracy (12 per cent female urban illiteracy, 14.2 per cent rural female) (*Comercio* 27 August 1994). Another article notes physical growth differences between rural and urban women due to malnutrition. The differences in literacy, height and numbers of children are drawn into a productionist discourse looking at 'productivity throughout the lifecycle' (*Universo*, 9 March 1994). Implicit are judgements about the relative contributions made to national development by different groups of women; the disempowering effects of communal and national masculine hegemony on women are not covered.[13]

It is when women are apparently breaking with a role of domestic-based assistance that great concern is expressed. Reports on high rates of female Sierran agricultural work due to male outmigration link with themes about *campesina* women becoming 'masculinized' (*Comercio* 24 October 1993). One editorial in the national daily *Comercio* raised the issue of the 'feminization of agriculture and of rural poverty', where it was said that women were taking on the work-parties (*faenas*) that were 'nearly always hard and back-breaking' (*rudas y agobiantes*) (*Comercio* 3 November 1993). But, it continued, women were not sufficient in numbers to replace the 'men who roam the cities urged on by deception'.

The conclusion drawn from this and other articles is that modernizing agriculture is a priority goal: the failure to achieve modern agriculture results in the de-feminization of peasant women and the loss of male roles. One article concludes: 'Now what is required is a process of planned and funded modernization so that agriculture does not die . . . since agriculture is called to become one of the pillars of development' (*Comercio*, 3 November 1993). In most articles, large-scale planned interventions are proposed, in which peasant women's extensive experience and knowledge are glossed over. Women's participation in agriculture is presented as anti-modern, and by implication, anti-national. Another article argues that there is urgent need for reform 'since it's a question of convenience for the whole country' (*Comercio* 10 June 1994).

In summary, poorer women, especially in the Sierra, are represented as embedded in certain activities and places which do not fit with modernist development notions. Women are expected to take part in certain, often implicitly urban, activities or service economies; otherwise they are a 'problem' for national development. The discursive identification of rural women goes beyond the (highly problematic) aim of 'integrating women into development', but rather pinpoints *campesina* women as figures whose characteristics and behaviours not only prevent their integration into 'national life' but also, crucially, hold back the development of the rest of the country. Moreover, it is often the poorer, black and indigenous women of Ecuador who bear the costs of development, despite their supposedly beneficial 'integration' (Moser 1993a; Conger Lind 1992b).

Focusing on representations of Ecuadorean femininities, this section has suggested that discourses of national progress are gendered and 'placed', (re-) creating geographies of identity, and practices embedded in place, which inform Ecuadoreans about their imagined community. Press and government discourse identify certain characteristics and behaviours in women, thereby 'locating' (Probyn 1991) women in certain places and social relations and constituting them in national development and narratives of belonging. Female figures appearing in official and media discourses are constituted as metonyms for 'national' femininity; women in elite households, Guayaquil and Quito stand in for officially sanctioned, largely hegemonic, national femininities. Andean rural women by contrast stand, like photographic negatives, for non-national femininities whose gendered and spatio-temporal positioning runs counter to national modernity. The ambivalence of women's nationhood is structured around place (also 'race' and class), highlighting the need to take these factors into account when examining gendered national identities.

GENDERED NATIONAL IDENTITIES AND DEMOCRACY

The massive mobilization of women (and other groups) in Latin America throughout the 1970s and 1980s through diverse social movements was one of the key features of the region during those years (Slater 1985; Escobar and Alvarez 1992). Organizing through neighbourhood organizations, village institutions, the Church, women's groups and human rights watchdogs, gendered subjects articulated a new sense of identity and a new way of *haciendo política*, 'doing politics' (e.g. Jelin 1990; Jaquette 1989). The articulation of new identities, despite its occurrence within the national space and often engaging with issues of democracy, citizenship and autonomy, has not however been much analysed from the perspective of national identities and other affiliations with place (but see Radcliffe 1993a, 1993b; Blondet 1990), and the focus here is specifically on the national aspect.

In previous sections, the highly gendered practices and representations –

via the technologies and ideologies of citizenship, family life, the media and development discourse – have been shown to position female gendered subjects in the national project. Restricted in their activities and representations in the nation, women – *qua* women – are interpellated into the nation through ambivalent and contradictory subjectivities which invite them into the imagined community of the nation through specific, almost stereotyped, relations. Called into the nation as 'modern' women, female subjects and embodiments are accorded only a marginal and contradictory place in the nation. Yet these relations – and the multiple, internally contradictory identities which these posit – are constantly changing, largely through the globalization of economies and cultures of Latin American countries, and due to the constant interactions between hegemonic and popular constructs of the 'nation'.

Two key spheres – politics and economics – shifted during the 1970s and 1980s in many countries, permitting the rearticulation of intersubjective relations and identities. The economic sphere became crisis-ridden both at the level of 'national' financial concerns (debt, repayments, restructuring of production) and at everyday levels (with high inflation, removal of price controls, contraction of employment opportunities), resulting in severe hardship among large sections of the population. In parallel with this, civilian governments were overturned throughout the continent by military regimes (almost exclusively authoritarian) whose restrictions of political parties and press freedoms, repression of dissent, and acts of violence against civilians became widespread, affecting people from diverse backgrounds and locations.

In this context, complicated by rapidly changing family structures and social relations,[14] women entered into sites in which they had not previously been engaged. Women's entry into the labour market was notable in many Latin American countries, with official female participation rates rising to between 12.9 per cent (Guatemala) and 25 per cent (Mexico) (Cubbitt 1995: 119). Official figures severely underestimate female work, however, especially in the interpersonal and familial labour markets of domestic service, agriculture and domestic manufacturing. Increased participation of women occurred in the small-scale service end of an increasingly complex informal sector, in domestic manufacturing, and in the *maquiladora* (trans-national manufacturing companies) export-processing factories. Work relations often called upon patterns of interaction and physical mobility distinct to that experienced previously by women; low-income women previously working in shanty-towns or villages in reproductive, domestic-based tasks turned to travelling to work, developing friendships and work-related relationships with other women (in gender-segregated workplaces). Earning an income through labour was also, except in some domestic service jobs, a new aspect of women's personal identities, whose previous labour contributions had often given them little access to cash (especially in semi-subsistence agricultural economies). Access to income, although severely restricted by falling family incomes and rapidly rising prices, also appears, in certain circumstances (e.g. Laurie 1995), to have offered women the possibility of developing a consumer identity.

Political restrictions also changed women's relations of intersubjectivity, often in ways unexpected by the military states and by the women themselves. As Jaquette notes, 'ironically, military authoritarian rule, which intentionally depoliticized men and restricted the rights of "citizens" had the unintended consequence of mobilizing marginal and normally apolitical women' (Jaquette 1989: 5). The mobilization of previously apolitical women (depoliticized through state gender ideologies, citizenship restrictions and social norms) occurred through the geographies and socialities in which they became increasingly involved. Whether through searches for disappeared children, in demonstrations for democratization, or in critical discussions of authoritarian relationships in small groups (organized by CEBS or emergent feminist networks), women's sense of community slowly became transformed. 'Out of such groups, a sense of solidarity and of cohesion as community members [arose]; the sense of isolation and alienation decreased' (Logan 1988: 353).

The transformation of women's 'imagined communities' and the concomitant rearticulation of collective identity and geographies of identity has often been attributed to the contradictions generated by the rapid socio-economic change experienced by the region from the 1970s (cf. Laclau and Mouffe 1985). State policies to repress political activism, and open economies up to international markets, generated contradictions in women's practices and identities due to 'gender-specific policies in consequence even if not in content' (Alvarez 1990: 260). The largest women's groups in the continent were found in Peru, Mexico and Brazil where different regimes had 'vigorously pursued state capitalist development strategies that brought women into education, production and politics as never before' (Alvarez 1990: 261). In Brazil (as well as Argentina and elsewhere), the military regime drew upon long-term ideologies of gender and nation which saw women as essentially apolitical, and hence less threatening to National Security (Alvarez 1990: 261): (non-military) masculinities were a threat, femininities were not. In Peru, despite a progressive military government and distinct state ideologies, similar contradictions were created. Faced with rising levels of economic hardship, violence and few channels of representation, Peruvian women organized collectively in spheres as diverse as cooking and reproductive tasks, Christian groups, and in negotiation with local organizations and government (Vargas 1991).

Demanding better services, a return to democracy and higher living standards were not the only facets to women's new positioning in the national space, throughout Latin America (Menendez Carrión 1988). Women's 'critical gender consciousness' (Saporta Sternbach et al. 1992: 211) developed as a function of the fact that their demands were 'tied [not] solely to survival but rather to constructions of identity and relations of power' (Conger Lind 1992b: 147). Challenging the hegemonic framings of their gendered place and identity meant for women challenging the constitution of their identities solely as household providers and passive recipients of aid. Women organized around 'practical' material issues but also to '[chip] away at hegemonic discourses about gender, development and politics' (Alvarez and Escobar 1992: 320). In women's case, it was precisely the space in numerous social movements which provided new relations of intersubjectivity and the formation of collectivities. However temporary and contradictory, these collectivities led to transformations in identity, including national identities. Transformation in women's sense of belonging and affiliation in communities, including the national community, occurred precisely as they discovered the power-saturated and highly gendered nature of the relations in which they had traditionally been ascribed, and through which they had been called to identify with the national community.

From holding (self-)representations of maternal, domestically inscribed identities, women voiced an affiliation with the neighbourhood (often via radical Catholic groups) and imagined a community of women, like themselves, struggling against male power as well as trying to put food on the table. These new 'cartographies of struggle' (Mohanty 1991) emerged in rural and urban areas. In Peru, urban women began to organize collectively around issues of provisioning for their families, yet their treatment by the state, municipalities and husbands prompted them to confront issues of domestic violence and 'change in the heart of patriarchal legitimacy' (Vargas 1991: 25). In converting to Protestant evangelicalism and taking an active role in new church groups, Colombian women renegotiated masculine roles in the family, resulting in the reduction in classic *machista* behaviour and the raising of household living standards for women and children (Brusco 1995). Peasant women in the Peruvian highlands were confronted with their marginalization from the state-defined community of 'peasants' during the Agrarian Reform, and articulated an 'imagined community' of female peasants, whose agendas and priorities brought them into conflict with confederations and the state (Crespi 1976). Women in the southern Andes created women-only federations on the basis of these new affiliations and communities (Radcliffe 1993a).

Such new urban and rural 'imagined communities' did not necessarily include references to stereotyped feminism, yet the fluid boundaries around groups and the flexible in-group relations meant that a gender-specific focus emerged widely (Saporta Sternbach et al. 1992; Vargas 1991). Middle-class, largely urban women, who previously contributed to Marxist politics were confronted by Leftist parties' dismissal of their concerns, resulting in their breaking from traditional political practice and ideology and the formation of self-identified feminist groups. Feminist critiques of society made connections between previously separate spheres, through which female exclusion had been so effectively articulated; the military dictatorships were rooted in authoritarian patriarchal relations, illustrating 'the "highest form" of patriarchal oppression' (Saporta Sternbach et al. 1992). 'Democracy in the country and in the home' became a slogan in Ecuador, Chile (where it originated) and throughout Latin America.

Via this critical gender identity – upturning the 'feminine consciousness' attributed to female subjects by the state and society – women's national identities were transformed. The alternative imagined communities posited by women in numerous social movements were communities whose very patterns of interrelations, discourse and geographies of identity challenged the basis on which national identities had been built in Latin America. As in other social movements, women's groups 'reconfigured the state's democratic discourse on rights, citizenship and the "nation"' (Alvarez and Escobar 1992: 327). Particularly in groups organizing around issues of human rights and disappearances women's emerging collective identities demonstrated these 're-imagined nations'. Where low-income women gained a forum or space in which to rearticulate their community, the critiques were severe. 'I was as sick of my husband's regime as I was of [President] Somoza's', said one participant at a feminist meeting (Saporta Sternbach *et al.* 1992: 228).

Women pursuing issues of political violence also linked 'personal' with 'state' powers, as in Guatemala and Argentina. The Guatemalan widows of civil war organized in 1988 a group, CONAVIGUA (a human rights organization, National Coordination of Widows of Guatemala), which mobilized some 9,000 women throughout the country. Uniting mothers of soldiers and of the disappeared (Schirmer 1993), the radical rearticulation of community and belonging is traced by CONAVIGUA women across historical dualisms of public/private, nation/family. One founder, Rosalina Tuyuc, sees a situation in which 'the State rules over men, men rule over women, and women rule over their children. . . . There's a pecking order from top to bottom. We've got to change this way of behaving inflicted upon us' (quoted in Kuppers 1994: 115). Guatemalan rural and indigenous women, previously isolated and uneducated (of 9,000 members only around 30 can read and write), came together to discuss the national constitution, laws and the Quincentenary (Schirmer 1993: 60), with a mixed response from the state. Members were invited into the National Dialogue where women were 'treated with respect, because of our painful experience' (quoted in Kuppers 1994: 115), yet also were abducted and killed (Schirmer 1993: 56–7).

The contradictory response to women's rearticulation of community also occurred in Argentina, again a country with a highly repressive regime in the 1980s. The Madres de la Plaza de Mayo (Mothers of the Plaza de Mayo) in Buenos Aires are renowned globally for expressing opposition to the military junta through selective and gendered use of public space (Fisher 1993). Articulating their opposition to the regime through a 'motherist' identity, the Madres initiated a mothers' community through their visits to police cells and barracks, drawing them into socio-spatial relations which both challenged and reproduced expected behaviours (for example, using 'feminine/maternal' public parks and tea houses as places to make plans and exchange information). The military government initially perceived this in terms of mothering roles, encouraged by the regime itself in its campaign to protect 'Western civilization'. Yet the regime reacted with violence and disappearances when the mother's role switched from passive carer to active, 'public' citizen (Radcliffe 1993b). The Madres of Argentina succeeded in organizing the earliest protest against military rule, however, and strongly shaped the agenda on the return to democracy, promoting a human rights bill to bring former military leaders to trial (Perelli 1994).

Although the main node around which women in social movements rearticulated an affiliation with place/community was gender, it was not the only one. Nor was it without its contradictions. Multi-faceted identities gave rise to shifting and temporary coalitions around 'identity', only to move on to other groups and identifications. Such was the case with lesbians (forming a new identity with regard to feminists, and to homosexual men [MacRae 1992; Saporta Sternbach *et al.* 1992]), and black women (forming the Red de Mujeres Negras de Latinoamerica y el Caribe [Latin American and Caribbean Black Women's Network] in 1992). In many other groups, women worked alongside men of similar class, ethnic or neighbourhood backgrounds in political activism which often raised new political agendas in the context of Latin American nationhoods (Escobar and Alvarez 1992).

The radical critiques of the male-empowering basis of nationalism have not, generally, been a route into powerful positions for women in Latin America. Despite their widespread mobilization and extensive political and discursive work, women remain generally disempowered and marginalized from formal spheres. Many women believe that they should retain autonomy from traditional political spheres, yet others have attempted to bring changes in practices and discourses to national and municipal bureaucracies (e.g. Alvarez 1990). What is perhaps the most significant, if contingent and unstable, result of women's participation in social movements has been the (complex and contradictory) rearticulation of affiliations to place (neighbourhood, nation, Latin America) and to collectivities (communities of women, mothers, black women, indigenous and *campesina* women, and so on) that women have initiated and built upon since the 1970s. Although women have historically been placed in an ambivalent and disempowering position in the imagined national community, national identities are *now* more clearly for many Latin American female subjects one of the 'competing discourses and practices [which] partially filter and redirect the political articulation of gender interests in Latin America' (Alvarez 1990: 267). In this sense then, new social movements have arguably had a profound, although often complex and unrecognized, impact on the national identities and affiliations with place and community among female subjects in Latin America.

Male and female subjects are offered distinct means and images through which to imagine themselves as part of the national community. Whether in leisure activities, in iconic statues or in national histories, men are envisioned strongly as part of – if not the initiators or defenders of – the nation, where masculine embodiments and qualities (racialized and sexualized in particular ways) are re-represented to men, through historiographies, monuments, media images, popular cultures and everyday practices, even through *machismo*. As noted by other writers on Latin America, masculine subjects (when sexuality and 'race' are appropriate) are clearly interpellated as national. By contrast, this chapter has shown how female subjects – even accounting for class, locational and 'racial' characteristics – are marginalized ambiguously from the national community. Their re-presentation as domestic, Marian figures, whose community of concern is the family, and perhaps the neighbourhood, means that they are more selectively and restrictedly drawn into the national community. Although at times becoming icons of Latin American nations, female subjects remain associated with the maternal attachments and private spaces through which women have experienced familial and wider power relations. The persistence and strength of these ways of imagining women, especially urban women of white/creole descent, have been discussed in relation to the Ecuadorean press, female icons, and forms of citizenship found in the region. The paradox of being both at the centre and at the margins (Rose 1993: 151) of the national imagined community simultaneously disempowers women and offers them the fragile possibility for re-imagining and re-positioning themselves.

The rise of social movements throughout the continent has arguably transformed the imagined communities of female subjects, and their affiliations with place. As women's sphere of activities and their social interactions were changed during periods of economic crisis and political repression, so the ways that they imagined themselves – and the way in which others imagined them – were subtly yet powerfully transformed. No longer were women simply domestic, maternal figures whose ambiguous social and citizenship status gave them no wider community. Suddenly, women were active and majority participants in diverse movements demanding changes to the state, to ways of governing (in parliament and in the home), to interpersonal relations, and to gender relations. These new communities – and the social relations through which new identities could be provisionally formed – were often contradictory, short-lived and shot through with other social divisions ('race', class, sexuality being the major fracture lines). Nevertheless, the persistence with which women made reference to the state, new intersubjective relations, and changing parameters of social practice and (self-)representation imply a major change in their identities and the processes of identity formation. Whether such transformations in self-identity, and community re-formations, profoundly challenge the 'official' nationalisms and widespread social practices through which women are marginalized, remains to be seen.

7

REMAKING THE NATION

Democracy and belonging

People in Ecuador, as in many other countries, are aware of the contradictions generated by power-saturated national projects, and claims on their collective identification. Ecuadoreans demonstrate a great awareness of the context for their lives when asked about national identities. Rather than take on board unquestioningly a nationalist agenda, they assess different elements of it in light of their own subjectivity and social practice. Ecuadorean identities are constituted in and through the extremely powerful nationalist ideological work done daily – to return to Stuart Hall's phrase with which we began this book – yet they are not constituted through it alone. Being part of self/other-awareness, national identities are part of a response by ordinary subjects to official ideologies and practices of nation-building. Such responses are contradictorily and simultaneously emotion-laden and thought-through, an ideological production by everyday people.

Belonging to a nation is taken for granted by most individuals for most of the time, not something 'submerged' awaiting mobilization by the state in crisis (cf. Giddens 1991). Given the de-centred social sites for the constitution of identities, and the fracturing of state attempts to fully co-ordinate national space-time, popular national identities cannot be considered as either sub-merged or transient. Although winning the World Cup, to give one example, may provide one transient moment of identification with the nation, it is not the only one. Other cultural practices and popular sites provide arenas for alternative expressions of national affiliation, which are differently expressed yet equally valid.

Such a conceptualization raises doubts about the approaches to national identity which phrase it largely in terms of ethnic-racial origins. For certain writers, the creation of a single ethnic-racial origin point for the nation is fundamental to national identity. The imagining of an ethnic-racial nation-hood entails a dual practice, on the one hand looking back (in a search for a, largely mythical, genealogy and ancestry), and on the other hand, looking to the future (with practices of controlling biological reproduction of the nation, by power effects worked through sexed, gendered and racialized bodies: Stasiulis & Yuval-Davis 1995). Rather than formulate national identity in this way (thereby marginalizing subalterns and legitimizing the racialized and gendered power effects of nations), our approach has been one that assumes from the start a multi-diasporic and fractured national society, de-centred by global, local and regional cultural expression. It is in such multi-racial and fractured societies that biopolitics are instituted, yet their effects and implications are, we argue, felt by subjects not defined by an ethnic/racial essence, but by cross-cutting multiple social identities.

In multi-diasporic nations, deeply structured around gender, class and 'race'/ethnicity, the perception and reception of the 'liturgies' (Mosse 1975) of nationalism vary considerably according to these social sites, as previous chapters have indicated for Latin America. In other words, the 'codes of reception' through which people apprehend 'their' nation are shaped by patterns of intersubjective relations formed along other axes of social differentiation. Certain codes of reception can be closely tied to the official versions of nationalism; for example, Ecuadoreans' constantly repeated negative comments against Peru and the Peruvians. Codes of reception among Afro-Ecuadoreans were more at odds with official nationalist discourse, as they imagined themselves to be a more widespread and numerous group than was recognized by others. Black Ecuadoreans, despite being marginalized and being seen as somehow not 'fully' national by official discourse and practice, were able, through the social support and collective identities of their own communities (in both rural and urban areas), to imagine themselves as constituting a significant segment of Ecuadorean society. Such re-imagining depended in part upon an imaginative geography, which positioned them more widely around the country than was generally acknowledged by governing elites.

Geographies of identities/correlative imaginaries

Geographies of identities refer to the multiple, and contingently fixed, places (and social collectivities embedded in place) with which – and in relation to which – individual and collective identities are formed and expressed. The juxtaposition of these places/peoples provides a means – used consciously and continuously by Latin American citizens, according to our research – to make statements about their affiliations to the nation. By contextualizing their understanding of the nation through reference to these geographies of identities, subjects express complex ideas about how they see the national community and their position in it.

While official nationalist histories tend to take the territorial outline of the country for granted (as do nationalist geographies), this contrasts with the contingency of other spaces of belonging expressed in 'popular' geographies of identity. The official project of nation-building may appropriate and/or re-circulate certain of these geographies of identity. For example, a creole elite geography of identity places Quito at the heart of Ecuador emotionally, politically and geographically. The centrality of Quito to Ecuadorean national identities is reproduced widely in official documents and discourse, as a key element of official imaginative geographies (see also Radcliffe 1996a). However, non-official geographies of identities mostly circulate outside the official project, and are frequently at odds with it. Geographies of identities are also further transformed by the changing political, economic and cultural context of the region, which continuously re-orients and fractures simple boundaries around particular geographies, a point illustrated with reference to pan-regional and neighbourhood examples.

The debate about the possibility of a regional Latin American identity represents one site where non-national affiliations are explored and expressed. The debate continues in the intellectual musings of television personalities, and in the mass support for neighbouring nations' football teams. A Latin American identity has historically rested upon understandings of shared languages, colonial experiences and religion. More recently, globalizing cultural practices (such as football and television) and their consumption by millions around the continent have provided other sites for such regional imagined communities. International economic and political relations and policies (exemplified by the highly commercialized production of television programmes and sports coverage, and international trade agreements) reinforce, and to a certain extent underpin, such imaginings. Yet this regional identity is currently abutted against the increasing Hispanicization of the United States, as well as rising rates of Protestantism within the region, and increasing contestation of the benign interpretations of the colonial legacy. All of these processes thereby fracture the boundaries between the United States and Latin America, between north and south, which have historically framed *latino* affiliations.

At a different level, 'local' claims on affiliation and belonging revolve around issues of land in rural areas while in urban centres, neighbourhood communities provide sites for expression of identity. In 'neighbourhood nationalism', subtle social and spatial boundaries are discursively deployed to express belonging. In Lima for example, class and local affiliations can be renegotiated through the appropriation and re-contextualization of terms associated with national belonging, such as citizenship. In one low-income Limeño neighbourhood then,

> All the 'neighbours' [*vecinos*] are 'citizens' [*ciudadanos*], but not all the citizens are neighbours. In practice 'neighbour' has become a term which synthesizes primarily the identity of the *poblador* [barrio dweller] as an inhabitant of the poorest zones of the city. A *poblador* guards his[/her] distance, and knows him[/her]self to be different to any citizen of a residential neighbourhood.
>
> (Tovar 1986, quoted in Degregori *et al.* 1986: 158)

Yet the rapidly declining standards of living for great numbers of urban Latin Americans, and the uneven consolidation of *barrio* housing, mean that the boundary between a *poblador* and other urban populations has been fractured and re-made. In Lima the emergence of a hybrid musical form, *chicha*, comprising a mixture of Andean and creole elements, reconfigured historic creole–Andean divisions. Additionally, 'provincial' forms of social interaction emerged which offered hybrid spaces and practices for the rearticulation of neigbourhood – and regional – affiliations and the redrawing of boundaries between residents in low-income and middle-class neighbourhoods. Of course, such rearticulations were framed by the state legislation governing the amount of airplay to be dedicated to 'folkloric' music (Turino 1991). Just as pan-Latin American identities were being challenged and transformed through globalization, so too localized identities in urban areas were articulated through rapidly developing ideas about the nation and its make-up. Such 'boundary changes' in the city or region offered the possibility for reorienting the 'trajectories of affiliation' through which subjectivities coalesced. By providing boundaries, defined socially and spatially, to the communities with which people identify, the geographies of identities provide for (fragile and contingent) spaces of belonging, sites for emotional bonding. These geographies are also constituted as forms of taken-for-granted sociality, a sociality whose relationship with feelings of national belonging is not automatically given. Such reiterative practice highlights the fact of national identities' 'proximity to everyday life' (Cohen 1994), as well as the daily work done on them.

While geographies of identities privilege place in the production of national identities, correlative imaginaries shift the emphasis to the subjective – metaphorically the 'heart of the nation' – and the huge emotional investments that are made in nations by people. However, as we have argued these investments are not innocent, but the product of ideological work that produces national identities explored in this volume. We suggested in our overview that the form of doubling between the self and the social character-istic of correlative imaginaries generates positionalities which are part of the ensemble of relations that give rise to national identities. These identifications are a central part of the experience of a national identity and are precisely those moments of national pride, or shame, sadness or elation, which – amid the fractured sites of identity formation – offer perhaps for only a fleeting period of time, a moment of centring in a de-centred world. This is, in part, why national identities are so powerful, from the neighbourhood to the national anthem, and why the consequences of these correlative imaginaries can move people to tears or violence.

These moments of centring amid the fractures of de-centred nations offer a sense of home and belonging which – although constructed within the time-space frame of the nation – can often be manipulated as outside, or beyond, the conventions of temporality and spatiality. This is a part of the imaginary of the identification between the self and the social within specific sites and contexts, many discussed in this book. What this suggests is a poetics of the nation in which the moments of correlative imaginaries give voice to the sentiments and emotions of nationals offering a 'space' of belonging from which to speak. These are, indeed, the seductions of national identifications which are available for reframing within nationalist discourses and therefore the generation of intolerant ethno-nationalisms, the consequences of which are violence and death. A sense of home and belonging does not offer only a benign trajectory. This romance belies the conditions of existence of national identities within the contradictions of late modernity.

Ideological work by nation-states and citizens

The nation-state makes considerable investments in the creation and the sustenance of a national identity. In Ecuador alone – a small and relatively impoverished country by Latin American standards (Corkhill and Cubbitt 1988) – the nation-state carries out (and reassesses) actions and organizations of national space-time, population and territory which it expects will result in greater expressions of belonging to the 'nation'. These include measures to organize the nation's space-time, providing a narrative of historical continuity for the national territory and points in time and space for the remembrance of key moments of that narrative. Through civic moments in schools, commemorations and monuments for certain dates, people and places (cf. Gillis 1994), and the writing of nationalist histories and geographies (cf. Hooson 1994), the space-time of the nation is homogenized and re-represented to citizens, as it is through official control over mapping and land-titling. In addition, the nation-state is engaged in the definition and governing of the national population, through specific ideologies of national racial formation including *mestizaje*, new indigenism and 'whitening'. By means of state gender ideologies and practices controlling the expression of sexuality and relationships, the nation-state shapes the reproduction of the national population, as well as using its policies of education and certain biopolitics (related particularly to children, conscripts and women) to influence citizens' subjectivities. Not least, the state is involved in activities oriented to the maintenance of its territory and sovereignty, by means of a military infrastructure, the marking and defence of borders, and international relations. In these numerous ways, the nation-state acts in ways which it hopes would mobilize and interpellate groups for the governing project. The weight of ideological work of nation-building and creating an imagined community can be seen in these diverse and manifold processes and their associated discourses. The state may, via these activities and representations, succeed in fixing certain key features on the 'horizon' of subjects' national identities, at least the sense and knowledge (even if contested and challenged) of being an Ecuadorean, a Colombian or whatever. As Billig (1995) reminds us, people do not tend to forget their national identity. In Ecuador, despite a lack of universal schooling, illiteracy and uneven access to the media, all subjects that we interviewed acknowledged that they were Ecuadorean.

Yet ideological work is not only being carried out by the state, although this has tended to be the focus of previous studies on national identities (e.g. Billig 1995). One theme which emerged in our research is that popular ways of articulating and re-expressing national identities are relatively structured, rather than being a process of passive ideological consumption. In other words, structured world-views affected Ecuadoreans' position on the nation, and resulted from considerable amounts of quotidian ideological work, positioning their subjectivities in relation to the national discourses on identity. Their national identity was not a 'false consciousness' – as suggested by cruder versions of Marxist theory – but an expression and discussion of their situatedness within the nation-state's ideologies and practices.

The national calls into play, among popular subjects, a diversity of evaluations, exclusions, and self-awareness. This means that not only are *nation-states* engaged in ideological work which reproduces and rearticulates nationhood but that individual and collective *subjects* are engaged in this process too, albeit from within the diverse and contradictory sites they occupy. In other words, the co-existence of diverse trajectories of affiliation necessarily draws subjects into a continuous process of restating their positionality *vis-à-vis* the nation. Whether this 'popular ideological work' is collective or individual, the national is an arena with which engagement is made. At this level then, the official discourses and practices constantly reiterated and reprogrammed by the nation-state do interpellate with citizens' multi-faceted identities.

Moreover, affiliations with places and peoples – whose realms and activities do not coincide with the nation – constantly mediate peoples' expression of national identities. Other people and places (which can be subnational or supranational) drew up and created feelings of belonging, around which national identities were reassessed. Such conditionality or qualifying of national identities was reiterated by reference to the sites, apart from the nation, through which subjectivities were constituted and expressed. De-centred hegemonic relations, in practice and discourse, through which subjects are constituted give rise to a multiplicity of sites in which ideological work is being done. The relation between subjectivity and identity, rather than being mediated only by class relations, or nationality, is instituted in the *interactions* of gender, class, ethnicity, age/generation, nation, location, religion and occupation.

However, communication of such quotidian ideological work by citizens is highly structured by differentiated power and access to means of expression. Language serves as the official medium of communication of the nation-state, and also the codified means through which senses of community and difference can be reproduced across the national territory. The power of communication often resides in media forms (newspapers, novels and other literatures), yet non-verbal forms of communicating ideas about community, nation and identity have, both in the past and today, provided other arenas for expressing ideas of community. Illiterates and those without access to an increasingly commoditized (and globalized) circuit of visual and print media do not have less to say in the discursive process of imagining the nation. But their comments and practices, through which ideas of nation and identity are negotiated and expressed, do not reach the reifying (and massifying) forms of newspapers, novels and the television screen. Despite some increase in popular subjects' access to diverse media (radio being the prime example in contemporary Latin America), an analysis of national identities has to go beyond an examination of these formalized and 'public' discourses.

As should be clear by now, our analysis of national identities in Latin America attempts to go beyond the view of identity – or indeed nation – as a natural, bounded, essential facet with cross-cultural references. Handler suggests, quite rightly, that the term 'identity' has too long underpinned a 'globally hegemonic nationalist ideology' and that for this reason, there is a need for another language (Handler 1994: 28). Handler seems to suggest that when expressing an identity, people have some bounded homogeneous notion of themselves in their own heads. Yet our work emphasizes the constant work of juxtaposing projections and imaginations of communities/places, and the fact of people's self-conscious distancing and/or affiliations with them. In other words, however much the external (often official, and largely massified) expressions of identity project an image of homogeneity (which Handler rightly criticizes), our going behind that projection reveals complex and contingent processes and identities. That Handler's main concern lies with the externalities of national identities is further demonstrated in his suggestion to adopt the term 'status' rather than identity (Segal and Handler 1994). National status, they argue, focuses attention on the assignment of different social privileges to various ranked groups, a move which has important political implications. Yet to assume that people's primary or overarching identity is with the nation, unaffected by other, crosscutting, trajectories of affiliation, is to reify – if by a different route – the very homogeneity and fixity attributed to the nation form in the first place.

Nationhood in the de-centred social

Once it is recognized that identities are provisional, contingent and relatively unfixed, then the constant mediation and questioning of the national projects has to be taken on board. The constellation of sites and relationships through which multiple identities are formed have been a continuous theme of the book. Thus, subjects are constituted through the discursive practices and power effects of gender, sexuality, 'race' and class, simultaneously as the investments in nation-building impinge upon their subjectivity. The process, as Desmond expresses it, involves the 'multiple ways in which each person is socially situated vis-a-vis a variety of trajectories of affiliation' (Desmond 1993: 104). The 'variety of trajectories' along which people may express affiliation – or crucially *difference* – have been outlined in the discussion of Latin America. 'Official versions' of *ecuadorianidad* are haphazardly taken up and also questioned persistently by people constituted differentially through 'race', class, gender, age/generation, location and sexuality. In the Ecuadorean case, the trajectory of indigenous ethnic affiliation has become more public and 'politicized' since the late 1980s with the mobilization around CONAIE. Additionally, Afro-Ecuadoreans are claiming their own affiliations and sense of difference from the jointly *indigenista* and *mestizaje*-dominated state project.

In part, the various trajectories of identity have been constituted and reproduced by means of the globalizing media, into which Latin American countries have long been inserted. The circulation of European, North American and other Latin American countries' media productions through-out the region offers no fixity to the national territory, as we discussed in Chapter 4. The attempted homogenization of the national time/space by the nation-state, through its regulation and discursive positioning of the visual media particularly, is offered with no possibility of closure.

The constitution of a single 'Other' through which to define identity has been shown to be problematic, due to the sheer variety of 'Others' through which national identity can be expressed. In Ecuador, national identities are expressed in relation to the 'Other' of Peru, but also, depending on context, *vis-à-vis* the United States, Colombia or Europe. In summary, a constellation of potential 'Others' defines more complex and context-specific subjectivities. Thus, despite broad correspondences between Afro-Ecuadoreans and Afro-Colombians in representation, social practice and positionality in the nation, to attribute similar identities to them as black subjects in highly racialized nationalisms, is to deny the complex contextual articulation of local, regional and class identities in each country.

National identities and modernity

As noted by several writers, the current form of the nation is deeply embedded in notions of modernity, and its Enlightenment correlates of progress, reason and individuality. The weight of modernist ideas of nationhood and identity is notable in Latin America, including Ecuador, a region profoundly engaged with Enlightenment discourse. The self-conscious discourses and practices through which the Enlightenment ideals have been realized in Latin America attest to the importance of modernity and progress to official national imaginings. Despite the developing countries being 'Others' for the rise of the West, the nation-form was taken up by the south as a positive, rational and modern political model (Chatterjee 1993).

Of course, as we mentioned in Chapter 1, the nation-form is closely tied into Latin America's own historical trajectory such that questions about the 'origin' of forms are lost in the complexities of eighteenth-century global interchange and influences. Moreover, the formation of national identities in metropolitan countries such as France and England was profoundly entangled with national identity formation in colonial – and early post-colonial – countries (Canny and Pagden 1987). The formation of national identities was also embedded within the historical construction of 'races' (Segal and Handler 1994: 3). The hybrid form of the nation is thereby steeped in relations of imperialism, colonialism and construction of 'races' (Segal and Handler 1994; Said 1994). Given the rise of nations within this context, issues of 'race' and ethnicity and nationality are not distinct variables but are interconnected facets of larger processes of identification, as Segal and Handler rightly argue; metropolitan national identity rested (and to a degree still does rest) upon notions of opposition to 'raced Others' (Segal and Handler 1994: 1). Ethnic homogeneity in the metropole was a grounding for achievement of nation-hood, while racialized 'Others' were denied nationality.

Given these global relations of difference, the nation became an ambivalent and contradictory form for those countries in the position of being 'Other' to the West, according to several writers (e.g. Bhabha 1990b; 1994). In settler societies, with their inherently multi-racial and socially-divided populations, the idea of nationhood immediately generated a contradiction between the ideal ethnically-homogeneous nation, and the reality (Stasiulis and Yuval-Davis 1995). Such contradictions generated marked (and often racialized) boundaries between the idea of a national collectivity (defining who were citizens, who were not) and wider society, compounded by metropolitan countries' concurrent attempt to define nations racially. In Latin American countries with large indigenous populations, although they were arguably not all settler societies, similar contradictions arose. With the West defining colonized regions as their racialized Other in order to create their own identity, Latin American countries were caught in a dilemma of rejecting such Otherization or aligning themselves – with all the complexities of their multi-ethnic populations – with a racialized nation-form. In Ecuador, as we saw in various chapters, nationhood was defined very largely on racialized terms, yet the continuing ambivalence about *how* to racialize the national community (to create a nation through *mestizaje, indigenismo* or whitening?) indicates the impossibility of resolving such contradictions within such a racialized dynamic. The embarrassed and self-knowing response of certain Ecuadorean elites to questions of national identity and 'race' attest to this ambivalence.

However, such ambivalence downplays the consistency and thoroughness with which the nation-state in Latin America has taken on board other modernist notions as the basis for planning, legislation and state practice. Spheres as diverse as development discourse, education policy, military structures, and gender relations have all been profoundly affected by Enlightenment narratives of progress, rationality and biopolitics, each with their own power effects. The goal of homogeneous spread of 'development' through the national territories and populations of the region has been a primary rationale for policy and politics in Latin America for several decades, in countries as diverse as Ecuador, Brazil and Peru. Yet it is not only in this official imaginative geography that the modernist discourses of nation have been expressed in Latin America. Modern biopolitics, which seek to transform the bodily practices and affiliations of nationals, have also been utilized in Latin America to create modern nations. Such biopolitics have been expressed in spheres as apparently diverse as military-run school programmes in Ecuador, leisure and sport activities in Argentina, and numerous educational activities throughout the continent. Gender relations and ideologies in Latin American states are also strongly informed by Enlightenment notions of male–female difference and incomparability, revealed in discourses about the need to 'modernize' a nation's women, and more subtly through the positioning of male and female citizens in the workplace, politics and the family.

Modernizing narratives which highlight the individual rational self as the basis for a wider society have been contested since the 1980s by a politics which highlights the fractured and de-centred nature of the social, and the existence of alternative collective communities. Such politics – articulated through certain of the region's highly diverse social movements – politicize the processes through which citizens are interpellated into the nation. In other words, while the self-conscious nature and selective interpellation of nationalist ideologies have continued, these new social movements have mobilized subjects on a wider basis and in relation to a distinct politics than that articulated previously. One such politics is articulated by the CONAIE Ecuadorean indigenous confederation, in its emphasis on the fluidity and contingency of indigenous identities, and the need to reclaim a collective identity distinct from official, *mestizaje*-informed notions of progress and individual self-hood. The modernist discourse of single affiliations of rational individuals is fractured by the Ecuadorean indigenous movement's insistence upon the persistence and reality of alternative ways of seeing 'communities', their 'development', and the socially determined nature of 'racial' categories and identities. In the aftermath of the indigenous uprising of 1990, indigenous groups and individuals noted how, despite the weight of hegemonic discourses on nation and identity, there were also subtle shifts in the positioning of 'indigenous' peoples in national narratives. Indigenous people received fewer examples of direct discriminatory behaviour in their everyday lives, and the media were more willing to utilize positive images and material from indigenous perspectives. The uprising of the self-declared Zapatista National Liberation Army, in Chiapas state, Mexico, in January 1994 exemplifies another alternative to the modernist nationhood. While expressing an identification with Mexico, Zapatista discourse revealed the process of renegotiation over national identities through the juxtaposition of other affiliation to communities and places, beyond the Mexican nation defined (so ambivalently and contradictorily) as rooted in pre-Colombian origins and *mestizo* progress (also Burbach 1994).

Yet the politics of CONAIE and the Zapatistas cannot be reduced to a single moment. In both movements, relations of intersubjectivity around gender, citizenship, issues of difference (gender, ethnicities and the north–south divide), and around places (land titles, regional economies, households, international relations) are also being renegotiated and manoeuvred. In other words, while 'race' is one (perhaps the predominant) trajectory for contesting official affiliations to the nation in the contemporary Mexican and Ecuadorean examples, it is not the only one (cf. Mallon 1995). Culturalist arguments being used by CONAIE and the Zapatistas indicate that even what appear at first light to be 'indigenous' movements are actually engaged in the reconfiguring of indigenous identities around cultural issues rather than 'racial' essences. Such diversity of agendas and avenues for affiliation and coalition-building is what marks a profound alternative discourse to modernity and the nation-form that has historically prevailed in the region.

Yet the power of these alternative discourses to displace and reformulate prevailing practice and discourse is highly contingent and localized. So despite the emergence of alternative modernities, the nation-state and hegemonic discourses of national affiliations continue to offer positionalities to women, to 'non-national' sexualities, and to those marked as racially/ethnically 'Other', that are highly marginalizing and contradictory avenues for incorporation into the nation. The projects being articulated by various social movements around the region – including the Zapatistas, the Ecuadorean and other indigenous movements, as well as female and homosexual politics – represent an opportunity for a wider rearticulation of the modernist national narrative, in order to include those previously marginalized groups in nations. It remains to be seen whether political will and social flexibility can be mobilized to bring about the necessary transformations.

Democracy, nation and belonging

The increased disillusionment of ordinary people with the groups who govern them, and claim to represent them, is one feature of Latin America in the mid-1990s. Despite the redemocratization process of the 1980s which promised profound changes after years of authoritarian and exclusionary rule, the promise was not seen to be fulfilled. After electing politicians and presidents to run open, accountable and non-authoritarian regimes, citizens in many countries were faced with widespread financial and political corruption, various exclusionary practices, and, at times, the resumption of arbitrary and unaccountable rule. The situation was worsened throughout the region by financial insecurity, leading to economic restructuring and hardship. It is in this context that the CONAIE group in Ecuador and the Mexican Zapatistas are attempting to rearticulate notions of democracy, society and progress. Ironically, various social movements and intellectual projects were articulated for Latin America, taking as their basis a firm representative democracy, at a time when disillusionment and disenchantment with traditional politicians was at an historic peak. From within a general pessimism there was the 'deepening' of democracy in order to consolidate the radical potential of the new social movements. New ways of 'doing politics' in the 1980s and early 1990s were envisioned as ideally occurring under the umbrella of a truly redemocratized political system open to a variety of social actors (see, among others, Escobar and Alvarez 1992; Esteva 1987; Yúdice et al. 1992). In this framework, pessimism about 'traditional' politics generates varying effects on popular senses of nationhood.

In Brazil, low-income urban-dwellers in Salvador expressed very negative views about government and politicians, revealing profound alienation from the corrupt political parties in power (McCallum 1994). Bahiana respondents highlighted the class differences between themselves and the thieving 'big people' who, they thought, would inevitably continue to exploit the poor and thereby reinforce their feelings of hopelessness. Through these local codes of reception mediating state actions and discourse, low-income groups inverted official beliefs in civilization and modernity and thereby reconfigured the nation. In one low-income Bahiana neighbourhood, national football success and carnival provided brief respite to daily misery as well as a site for national belonging. In this case, football, street festivals and carnival provided sites in popular culture through which individual insertion into the space-time of the nation could be renegotiated. The correlative imaginary of these localized subjects was therefore constituted on the basis of distancing the state, and foregrounding other sites for constituting collective intersubjectivities as the 'nation'.

In Ecuador, despite the rise of opposition movements around gender and indigenous issues, the degree of alienation from the state and governing groups would not appear to be as deep as in Brazil. Ecuadoreans identified a variety of different powerful groups, refusing to divide the politicians from the 'people' so sharply, as appears to be the case in Brazil. Rather, individual power was identified alongside the powers of the military, and the Catholic Church. Political conflict between social groups was not generally regarded as a significant issue, and when asked about the groups who had power in the country, the diversity (and unpredictability) of responses was staggering. People from all different class and racial groups identified different types of power-holders, from the office-holders at village level, through to municipal, departmental and national administrations. In a personal context, although many women claimed to have some power only in the home, generally individuals attributed some power to themselves, whether in the workplace, in a ritual or community role, or via moral and social precepts. When asked about the racialization of power, blacks were surprisingly eager to attribute power to themselves, despite their extremely limited representation in formal positions. In the provincial and largely indigenous district of Cotopaxi, just under half of respondents (49 per cent) claimed that the indigenous had most power in contemporary Ecuador (perhaps a reflection of the 1990 Uprising in which Cotopaxi played a significant role).[1] The correlative imaginary of Ecuadoreans would therefore appear to be one in which the de-centring of power, through formal and informal means, is significant in national self-imaginings. Ecuadoreans talk in terms of a de-centred national narrative, a narrative that the nation-state itself is constantly struggling to overcome. Individual subjectivities around nationhood take on board the issue of state presence without great alienation, yet the de-centring and strong region-alization which has long characterized Ecuador historically continues to circulate.

Yet even for Ecuadoreans, there is a minimum sign which would appear to provide consensus, and that is the Otherization of Peru. In Ecuador, the persistence with which Ecuadoreans use the 'Other' of Peru, defined socially and territorially, as the mirror for national expression exemplifies the use of an 'Other' around which to define 'nation'. Even if people forget why Peru is seen as the 'threat' or limit to Ecuadorean nationhood, subjects are at least minimally interpellated by this representation of themselves. The Peruvian 'Other' is a social cement, tying together the highly fractured and de-centred society that is contemporary Ecuador. The mobilization of self-identified indigenous people as combatants in the border dispute with Peru in January 1995 suggested that the boundary between the two countries – overlain as it is with official and popular imaginative geographies – transcends ethnic identities, and regional affiliations. Yet such a fragile identity around the 'logoization' of the Rio Protocol line, and the nationalist agendas which go with it, depends upon constant 'memory work' (Gillis 1994: 3), through which the national narrative can be tentatively stitched together (cf. Boyarin 1994).

By positioning themselves – however contingently and resistantly – *vis-à-vis* communities and places, subjects do 'memory work' about their national identities as the juxtaposition of diverse communities/places with the nation has to be organized. There are a multiplicity of ways through which the 'nation', even such a small country as Ecuador, is remembered and affiliations with it expressed. National imagined communities are, to repeat a continuing theme in the book, not only in people's heads – in the imaginations of a nation's citizens, but are projected and articulated through the media, education, cartography and educational practice – and also embodied and practised. These communities are found in everyday life in the practices which are taken for granted within contextualized socio-spatial relations, which remind people of their nationhood (cf. Billig 1995). While Gillis rightly highlights the public nature of commemoration, utilized to bind together nations as much in forgetting as in remembering (Gillis 1994: 7), our work on Ecuador suggests that individual processes of memory and its loss also constitute affiliations with the nation. In terms of the politics of national identities, it is a question of forging a democratic sense of belonging built on constantly changing patterns of diversity and hybridity, which together form the basis for the inclusionary project of making the nation.

NOTES

1 IMAGING THE NATION: RETHINKING NATIONAL IDENTITIES

1 Anderson defined the nation as imagined community 'because the members of even the smallest nation will never know most of their fellow-members, meet them, or even hear of them, yet in the minds of each lives the image of their communion' (Anderson 1991: 6). The nation for him is limited and sovereign.

2 However, several writers have noted that popular sentiments of national identity are not necessary 'force-fed' by the elites (Taylor 1989: 195). In Europe, nationalism's attraction may lie in offering an 'escape from the consequences of industrialization' particularly felt by the subaltern classes (Mosse 1975: 6).

3 Giddens (1991: 250) defines the nation as the unification of administration and power over a specific territory, emphasizing the infrastructure of the nation-state rather than its meaning for citizens. Smith defines the nation as 'a named human population sharing an historic territory, common myths and historical memories, a mass, public culture, a common economy, and common legal rights and duties for all members' (A. Smith 1991: 14).

4 The concept is broader than 'popular geopolitics' (Sharp 1993), the latter dealing with popular feelings about nations without questioning its configuration.

3 ECUADOR: MAKING THE NATION

1 Authors' interview with Education Minister, Quito, April 1994.

2 Interview with an official, Education Ministry, Quito, 1994.

3 Interview with Education Minister, Quito, April 1994.

4 Ibid.

5 The *Audiencia* is an administrative division used in the Spanish American colonies. Originally under the remit of Lima, Quito and its hinterland became an *Audiencia* in 1563, then responsibility moved from Lima to Bogotá in 1740.

6 Interview with Gen. Andrade, Director of the Instituto de Altos Estudios Nacionales, Quito, 1994.

7 Interview with Gen. Paco Moncayo, Army General Inspector, Quito, 1994.

8 Juan León Mera is mentioned frequently by our respondents, who name him the most famous Ecuadorean writer.

9 The chorus, loosely translated, is 'Save oh *Patria*, a thousand times! Oh *patria*!/ Glory to you! Now your breast shelters/Delight and peace, and your radiant brow/ More than the sun we contemplate shining'.

10 Interview with former state-employed geographer, 1994.

11 Interview with Director of Instituto Geográfico Militar, Geography Section, Quito, 1994.

12 Interview with educationalist, 1994.

13 Information about Francisco Terán was kindly supplied by his family, following his death in 1990. We are very grateful to the Terán family, especially Raul Terán King.

14 Interview with IGM Director of Geographical Information, Quito, 1994.

15 Figures from 1970s, Quintero and Silva (1991: 231); from 1990s, interview with Gen. Andrade, Quito, 1994.

16 Interview with Gen. Andrade, Quito, 1994.

17 Face colour (a gloss for race) was noted in passports until the late 1980s, but only a wealthy minority would have been affected. Officials dealing with passport applications would make a racial designation based on characteristics visible in the applicant's photo.

18 Such a concern is not just recent: the military-led 'Juliana revolution' of 1925 aimed to dignify 'the indigenous race' (Quintero and Silva 1991: 227).

19 Interview with Army General, Quito, 1994.

20 He said, 'At this moment there is no racism in the armed forces . . . no different treatment by race. Yes, we have a lovely mixture because we have indians, blacks, mestizos, and between ourselves it makes no difference.' Interview, Quito, 1994.

21 Interview with army spokesman, Quito, 1994.

22 Interview with IAEN spokesman, 1994.

23 The indigenous uprising of June 1990 mobilized peasants and indigenous populations throughout the country for some six days, involving road blocks, kidnapping of state officials, and mediation between indigenous and the state by key Catholic Church members (see Zamosc 1994; Almeida *et al.* 1992).

24 Interview with the Ecuadorean armed forces public relations officer, Quito, April 1994.

25 Ibid.

26 A further 8 per cent have further education, 0.4 per cent have postgraduate training, 1.2 per cent are literate (without schooling) and 9.8 per cent said they had no formal education, leaving 3.8 per cent for whom no information was available (Ecuador 1990).

27 The programme involved NGOs in adult popular education, resulting in an (under-funded) Permanent Popular Education department in the Ministry of Education. Interview with independent education consultant, Quito, 1994.

28 Bilingual education policies have been in force in Ecuador since 1985, and are similar to policies in other Andean countries such as Peru and Bolivia where the high percentage of non-Spanish speaking populations prompted an alternative educational system (de Vries 1988). Directed mainly at primary education and at Quichua monolingual children, the Educación Bilingüe Intercultural (EBI: Intercultural Bilingual Education) project was carried to seventy-four schools in eight provinces by 1986–7 (de Vries 1988).

29 Interviews with the Education Minister and an educational consultant, Quito, April 1994.

30 The Central Bank has played a long and significant role in the collection and display of national artefacts and cultures since its founding. It still runs a publishing house, which publishes books on culture and history.

31 The two exceptions to this are a display on coastal *cholos*, showing boats and nets, and photos about firework makers in a display emphasizing the festival-driven nature of firework production.

32 On the ground floor, large maps are displayed showing the environments and economic activities associated with each region. Next to these maps is a photo-mosaic in the shape of the country, with individuals (men, women, adults and children) of differing ethnic and racial groups. Also shown is a large physical relief map with the Rio Protocol line, and neighbouring countries shaded in yellow, under the heading 'Ecuador – ecological miracle'.

33 Interview with Director of INPC and officials, April 1994 (cf. INPC 1989; Anderson 1991). It is also notable that since the 1960s, the IGM's history commission has

been involved in the organization of museums, as well as the conservation of archaeological remains and historic monuments.
34 Interview with historian and adviser to museum, Quito, 1994.

5 NATIONALIZED PLACES? THE GEOGRAPHIES OF IDENTITY

1 It thus goes beyond an analysis of relationships between geography as an academic discipline and national identity(e.g. Escolar *et al.* 1994; Hooson 1994; Orlove 1993).
2 Such rural–urban distinctions and imaginative geographies are frequently gendered (Radcliffe 1996b).
3 We have no information about popular imaginations.
4 Of the twenty proposals for roads and colonization of the region between 1860 and 1920, none was realized.
5 Interviews nos 181 and 217.
6 In terms of the highly diverse literary canon, there is some emphasis on the 'indians' of the Sierra, and the Afro-Ecuadoreans on the northern coast.
7 'Only in peripheral areas to the south of the country did the old ethnic groupings or *ayllus* maintain still a certain relevance as group identifiers' (Albó 1987: 38).
8 The Galapagos reappeared as the most beautiful place for him, and the most ugly the Oriente, the region where he lived.
9 One picture shows two independence heros (Rosa Sarate, Nicolas de la Peña) being executed, the other shows an earthquake in Quito in March 1852.
10 Only 8 per cent in Cotopaxi replied that the whole country was beautiful. In Cotopaxi, people were reluctant to identify displeasing national sites, preferring to say that no place was ugly (37 per cent) or that they did not know (22 per cent).
11 Degregori *et al.* (1986) sense that if asked about differences at a national level, replies would have been distinct. One respondent hints at this when explaining the rise of *Sendero Luminoso* (Shining Path), the Maoist guerrilla group, in terms of *serranos*' child-like and passive nature.
12 In other countries too, conscription rates are down, as in Argentina where in 1983–9 draft declined by 15 per cent (Pion-Berlin 1991), or in Brazil where draft evaders represented one-fifth of total numbers liable (Rouquié 1987: 64).
13 Exchange rate in April 1994. If the fine is not paid, interest is levied on it until the conscript is 25 years old.
14 Protestant, Catholic and Jehovah's Witness churches are found but do not generate much social activity.
15 Much of the information in this section was collected during a visit and interviews with OPIP representatives in Puyo, April 1994.
16 Interview with OPIP cartographer, Puyo, 1994.
17 Waves of colonization into the Amazon region began in the 19th century, with the arrival of *mestizos* from the Andes (Zárate 1993). From the 1950s, state-directed and semi-directed colonization schemes increased numbers of immigrants (E. Salazar 1981).
18 Interview with OPIP cartographer, Puyo, 1994.
19 Interview no. 014, 1994.

6 GENDER AND NATIONAL IDENTITIES: MASCULINITIES, FEMININITIES AND POWER

1 However, in the 'West' there has arguably been a shift from private to public patriarchies (Pateman 1988; Walby 1990).
2 The meanings generated by this image are not stable. Stories from Brazil's northeast (the 'blackest' region) told how Xuxa made a pact with the devil, and how a

girl was killed by one Xuxa record. Adoption of Xuxa features by male transvestites emphasize the transgressive possibilities inherent in Xuxa's fundamentally conservative imagery.

3 Illustrated in discussions of numerous nationalist monuments to the Unknown Soldier (cf. Johnson 1995; Anderson 1991).

4 The instability of meaning in the *gauchos* is visible however in the anarchist utilization of this figure as an inspiration for social revolution (Archetti 1994d: 9).

5 Interview no. 246, 1993.

6 Cases of murder of male homosexuals are recorded in *En Directo* magazine (no. 5, April 1992), and Conger Lind (1992b).

7 Press and government reports are oriented to, and largely produced by, elite, 'white' urban classes, although the assumed homogeneous readership is in effect divided along regional lines (Guayaquil, Quito), and by class (with rapidly rising rates of literacy among all social groups).

8 National development plans and government agency reports from 1979 to 1994, and national newspapers of 1992–4 provide the empirical material for this examination.

9 The women are Rosita Campuzano, Rosa Zárate, Manuela Espejo, Manuela Garaycoa, Dolores Veintemilla, Marieta de Veintemilla; see Radcliffe (1996b).

10 The 1980–4 National Development Plan included a subprogramme for women and youth in its popular participation section (Ecuador 1984; cf. CEIS 1988).

11 However, the representation of Guatemalan indigenous women is embedded in racisms and male power which marginalize these women in many political (Kuppers 1994: 111–15; Schirmer 1993).

12 If such a figure is correct, then rates of female participation are equivalent to Sub-Saharan Africa. We are grateful to Lucy Jarosz for pointing this out.

13 Male and female rates of education, whether urban or rural, do not differ greatly. Nationally, 51.9 per cent of men have a primary education, compared with 49.9 per cent of women; 77.6 per cent of rural men have primary or secondary education, as do 72.4 per cent of rural women; 52 per cent of rural women over 15 years have four or more children, compared with 35 per cent of urban women. But there are more urban than rural children overall, and more rural women than urban have *no* children (ratio of five rural to three urban).

14 This is due to rising numbers of female-headed households, new legal provisions concerning family relations, and the influence of secular moralities on personal relationships.

7 REMAKING THE NATION: DEMOCRACY AND BELONGING

1 One-quarter (24 per cent) of Cotopaxi residents identified the whites/creoles as the most powerful racial group, and 15 per cent the *mestizos*.

BIBLIOGRAPHY

Abán Gómez, E., Andrango, A. and Bustamente, T. (comps) (1993) *Los Indios y el Estado-Pais Pluriculturalidad y Multi-Ethnicidad en el Ecuador: Contribuciónes al debate*. Quito: Abya-Yala.

Abercrombie, T. (1991) 'To be indian, to be Bolivian', in Urban, G. and Sherzer, J. (eds) *Nation States and Indians*. Austin: Texas University Press.

Abram, M. (1992) *Lengua, Cultura e Identidad: el Proyecto EBI 1985–1990*. Quito: EBI/Abya-Yala.

Adam, B. (1994) *Timewatch: The Social Analysis of Time*. Cambridge: Polity.

Albó, X. (1987) 'Formación y evolución de lo aymara en el espacio y el tiempo', in *Coloquio Región y Estado en Los Andes*. Cuzco, Peru: Centro Bartolomé de las Casas.

—— (1993) 'Our identity starting from pluralism in the base', in Beverley, J and Oviedo, J (eds) *The Postmodernism Debate in Latin America*, special issue of *Boundary 2*. 20 (3): 18–33.

Albornoz, P. O. (1971) *Las Luchas Indigenas en el Ecuador*. Quito: Libris.

Allen, C. (1988) *The Hold Life Has: Coça and Identity in an Andean Community*. Washington, DC: Smithsonian Institute Paper.

Allpanchis (1987) 'Lengua, nación y mundo andino'. *Allpanchis* 29–30 (19).

Almeida, I., Almeida, T. and Bustamente, S. (comps) (1992) *Indios: Una Reflexión Sobre el Levantamiento Indígena de 1990*. Quito: ILDIS/Abya-Yala.

AlSayyad, N. (ed.) (1992) *Forms of Dominance: On the Architecture and Urbanism of the Colonial Enterprise*. Aldershot: Avebury.

Altamirano, T. (1980) 'Regional commitment and political involvement amongst migrants in Peru: the case of regional associations', PhD thesis, University of Durham.

Althusser, L. (1971) *Lenin, Philosophy and Other Essays*, trans. B. Brewster. London: New Left Books.

Alvarez, S. (1990) *Engendering Democracy in Brazil*. Princeton, NJ: Princeton University Press.

Alvarez, S. and Escobar, A. (1992) 'Conclusion: theoretical and political horizons of change in Latin American social movements', in Escobar, A. and Alvarez, S. (eds) *The Making of Social Movements in Latin America*. Boulder, CO: Westview Press.

Andean Mission of the United Nations (1960) *Integration of the Native Indian Population in Ecuador*, (no place of publication given): Andean Mission/Government of Ecuador.

Anderson, B. (1991) *Imagined Communities: Reflections on the Origin and Spread of Nationalism*, 2nd edn. London: Verso.

Anthias, F. and Yuval-Davis, N. (1992) *Racialized Boundaries: Race, Nation, Gender, Colour and Class and the Anti-Racist Struggle*. London: Routledge.

Appadurai, A. (1986) *The Social Life of Things*. Cambridge: Cambridge University Press.

—— (1990) 'Disjuncture and difference in the global culture economy'. *Public Culture* 2 (2): 1–24.

Archetti, E.P. (ed.) (1994a) *Exploring the Written: Anthropology and the Multiplicity of Writing*. Oslo: Scandinavian University Press.

—— (1994b) 'Models of masculinity in the poetics of the Argentine tango', in Archetti, E. P. (ed.) *Exploring the Written: Anthropology and the Multiplicity of Writing*. Oslo: Scandinavian University Press.

—— (1994c) 'Masculinity and football: the formation of national identity in Argentina', in Guilianotti, R. and Williams, J. (eds) *Game Without Frontiers: Football, Identity and Modernity*. Aldershot: Arena.

—— (1994d) 'Nationalism, football and polo: tradition and creolization in the making of modern Argentina', Paper given at the 'Locating Cultural Creativity' conference, University of Copenhagen, October.

—— (1994e) 'Estílo y virtudes masculinas en *El Gráfico*: la creación del imaginacion del futbol argentino', Paper given at the 48 Congreso Mundial de Americanistas, Stockholm, July.

Austin (1994) 'The Austin Memorandum on the Reform of Art. 27, and its impact upon the urbanization of the *ejido* in Mexico', *Bulletin of Latin American Research* 13 (3): 327–36.

Balibar, E. (1990) 'The nation form: history and ideology', *Review* 13 (3) 329–61.

Balibar, E. and Wallerstein, I. (1991) *Race, Nation and Class: Ambiguous Identities*, London: Verso.

Bastian, J. P. (1986) 'Protestantismo popular y política en Guatemala y Nicaragua', *Revista Mexicana de Sociología* 3: 181–99.

—— (1990) *Historia del Protestantismo en América Latina*, Mexico City: Casa Unida de Publicaciones.

—— (1993) 'The metamorphosis of Latin American protestant groups: A sociohistorical perspective', *Latin American Research Review* 28 (2): 33–61.

Bebbington, A. and Ramon, G. (comps) (1992) *Actores de una Década Ganada: Tribus, Comunidades y Campesinos en la Modernidad*, Quito: COMUNIDEC.

Bell, D. (1991) 'Insignificant others: lesbian and gay geographies', *Area* 23: 323–9.

Bendix, R. (1964) *Nation Building and Citizenship*, New York: Wiley.

Berdichensky, B. (1986) 'Del indigenismo a la indianidad y el surgimiento de una ideologia indígena en andino américa', *América Indígena* XLV1 (oct–dic): 643–55.

Beverley, J. and Oviedo, J. (eds) (1993) *The Postmodernist Debate in Latin America*, special issue of *Boundary 2*, 20 (3), Durham, NC: Duke University Press.

Bhabha, H. K. (1990a) 'Dissemi-Nation: time, narrative and the margins of the modern nation', in Bhabha, H. K. (ed.) *Nation and Narration*, London: Routledge.

—— (1990b) 'Introduction: narrating the nation', in Bhabha, H. K. (ed.) *Nation and Narration*, London: Routledge.

—— (1994) *The Location of Culture*, London: Routledge.

Biersack, A. (1989) 'Local knowledge, local history: Geertz and beyond', in Hunt, L. (ed.) *The New Cultural History*, Berkeley: University of California Press.

Billig, M. (1995) *Banal Nationalism*, London: Routledge.

Binney, J. (1994) 'Embodying the nation', Paper given at the American Association of Geographers' Conference, San Francisco, March.

Bird, J., Curtis, B., Putnam, T., Robertson, G. and Tickner, L. (eds) (1993) *Mapping the Futures*, London: Routledge.

Blanchard, P. (1992) *Slavery and Abolition in Early Republican Peru*, Wilmington, DE: SR Books.

Blaut, J. (1987) *The National Question: Decolonizing the Theory of Nationalism*, London: Zed Press.

Blondet, C. (1990) 'Establishing an identity: women settlers in a poor Lima neighbourhood', in Jelin, E. (ed.) *Women and Social Change in Latin America*, London: UNRISD/Zed.

Blunt, A. and Rose, G. (eds) (1994) *Writing Women and Space: Colonial and Postcolonial Geographies*, London: Guildford Press.

Boff, L. and Boff, C. (1987) *Iglesia: carisma y poder. Ensayos de eclesiología militante*, Santander, Spain: Sal Terrae.

Bollinger, W. and Lund, D. M. (1982) 'Minority Oppression: towards analyses that clarify and strategies that liberate', *Latin American Perspectives* 9 (2): 2–28.

Bondi, E. (1993) 'Locating identity politics', in Keith, M. and Pile, S. (eds) *Place and the Politics of Identity*, London: Routledge.

Bourdieu, P. (1977) *Outline of a Theory of Practice*, Cambridge: Cambridge University Press.

Bourque, N. (1993) 'The power to use and the power to change: saints' cults in a Quichua village in the Central Ecuadorean Highlands', in Rostas, S. and Droogers, A. (eds) *The Popular Use of Popular Religion*, Amsterdam: CEDLA.

Bourque, S. (1989) 'Gender and the state: perspectives from Latin America', in Charlton, S., Everett, J. and Staudt, K. (eds) *Women, the State and Development*, Albany: State University of New York.

Bourricaud, F. (1975) 'Indian, mestizo and cholo as symbols of the Peruvian system of stratification', in Glazer, N. and Moynihan, D. (eds) *Ethnicity: Theory and Experience*, Cambridge, MA: Harvard University Press.

Bowman, G. (1994) ' "A country of words": conceiving the Palestinian nation from the position of exile', in Laclau, E. (ed.) *The Making of Political Identities*, London: Verso.

Boyarin, J. (ed.) (1994) *Remapping Memory: The Politics of Time–Space*, Minneapolis/London: University of Minnesota Press.

Brading, D. (1991) *The First America: the Spanish Monarchy, Creole Patriots and the Liberal State, 1492–1867*, Cambridge: Cambridge University Press.

Breckenridge, C. and Appadurai, A. (1989) 'On moving targets', *Public Culture* 2 (1): i–iv.

Breuilly, J. (1993) *Nationalism and the State*, 2nd edn, Manchester: Manchester University Press.

Bronstein, A. (1984) *The Triple Struggle*, London: WOW Campaign.

Brow, J. (1990) 'Notes on community, hegemony and the uses of the past', *Anthropology Quarterly* XV (3): 1–6.

Brunner, J. (1993) 'Notes on modernity and postmodernity in Latin American culture', in Beverley, J. and Oviedo, J. (eds) *The Postmodernist Debate in Latin America*, special issue of *Boundary 2*, 20 (3): 34–54.

Brusco, E. (1995) *The Reformation of Machismo: Evangelical Conversion and Gender in Colombia*, Austin: University of Texas Press.

Burbach, R. (1994) 'Roots of the postmodern rebellion in Chiapas', *New Left Review* 205: 113–24.

Burdick, J. (1992) 'Rethinking the study of social movements: the case of Christian Base Communities in urban Brazil', in Escobar, A. and Alvarez, S. E. (eds) *The Making of Social Movements in Latin America: Identity, Strategy and Democracy*, Boulder, CO: Westview Press.

CAAP (1981) *Comunidad Andina: Alternativas Políticas de Desarrollo*, Quito: CAAP.

Canclini García, N. (1993) *Transforming Modernity: Popular Culture in Mexico*, Austin: University of Texas Press.

Canny, N. and Pagden, A. (eds) (1987) *Colonial Identity in the Atlantic World, 1500–1800*, Princeton, NJ: Princeton University Press.

Cant, G. and Pawson, E. (1992) 'Indigenous land rights in Canada, Australia, and New Zealand', *Applied Geography* 12 (2).

Castañeda, J. G. (1994) *Utopia Unarmed: The Latin American Left After the Cold War*, New York: Random House.

CEIMME (1994) *Informe del Sector No-Gubernamental, Pais Ecuador, IV Conferencia Mundial de la Mujer: 'Acción para la igualdad, el Desarrollo y la Paz. Foro de ONGs'*, Quito: CEIMME.

CEIS (Centro Ecuatoriano de Investigaciones Sociales) (1988) 'Problems that concern women and their incorporation in development: the case of Ecuador', in Young, K. (ed.) *Women and Economic Development: Local, Regional and National Planning Strategies*, London/Paris: Berg/UNESCO.

de Certeau, M. (1984) *The Practice of Everyday Life*, Berkeley: University of California Press.

Chaney, E. (1979) *Supermadre: Women in Politics in Latin America*, Austin: Institute of Latin American Studies, University of Texas.

Chaney, E. and García Castro, M. (eds) (1989) *Muchachas No More: Household Workers in Latin America and the Caribbean*, Philadelphia, PA: Temple University Press.

Chartier, R. (1988) *Cultural History: Between Practices and Representation*, Ithaca, NY: Cornell University Press.

Chatterjee, P. (1993) *The Nation and its Fragments: Colonial and Post-Colonial Histories*, Princeton, NJ: Princeton University Press.

Chiriboga, M. (1983) 'El análisis de las formas tradicionales: el caso de Ecuador', *América Indígena* XLIII dic.: 37–99.

—— (ed.) (1986) *Movimientos Sociales en el Ecuador*, Quito: CLACSO.

—— (1988) *El Problema Agrario en el Ecuador*, Quito: ILDIS.

Chirif, A., García, P. and Chase Smith, R. (1991) *El Indígena y su Territorio Son un Solo*, Lima: Oxfam America/COICA.

Clark, A. K. (1994) 'Indians, the state and law: public works and the struggle to control labor in Liberal Ecuador', *Journal of Historical Sociology* 7 (1): 49–72.

Coad, M. (1992) 'Ecuador's new rulers plan gentle shock therapy', *Guardian* (London) 11 August.

Cohen, A. (1994) *Self-consciousness: An Alternative Anthropology of Identity*, London: Routledge.

Colley, L. (1986) 'Whose nation? Class and national consciousness in Britain, 1750–1830' *Past and Present* 33: 96–117.

CONAIE (1989) *Las Nacionalidades Indígenas en el Ecuador: Nuestro Proceso Organizativo*, 2nd edn, Quito: Abya-Yala.

Conger Lind, A. (1992a) 'Confronting development on their own terms: women's responses to national and global restructuring in Ecuador', Paper prepared for XVII LASA Congress, Los Angeles, 24–27 September.

—— (1992b) 'Power, gender and development: women's organizations and the politics of needs in Ecuador', in Escobar, A. and Alvarez, S. (eds) *The Making of Social Movements in Latin America*, Boulder, CO: Westview Press.

Connell, R. (1987) *Gender and Power*, Cambridge: Polity.

Cooke, P. (1990) *Back to the Future*, London: Allen & Unwin.

Corcoran-Nantes, Y. (1993) 'Female consciousness or feminist consciousness: women's consciousness raising in community-based struggles in Brazil', in Radcliffe, S. and Westwood, S. (eds) *'Viva': Women and Popular Protest in Latin America*, London: Routledge.

Corkhill, D. and Cubbitt, D. (1988) *Ecuador: Fragile Democracy*, London: Latin American Bureau.
Cortés, S. A. (1960) 'Breve reseña histórica del Instituto Geográfico Militar', in *Instituto Geográfico Militar 1928–1960: 32 años al servicio del la Patria*, Quito: IGM.
Crain, M. (1989) *Ritual, Memoria Popular y Proceso Político en la Sierra Ecuatoriana*, Quito: Abya-Yala.
—— (1990) 'The social construction of National Identity in Highland Ecuador', *Anthropology Quarterly* XV (3): 43–59.
Crespi, M. (1976) 'Mujeres campesinas como líderes sindicales: la falta de propriedad como calificación para puestos políticos', *Estudios Andinos* 5 (1): 151–71.
Cubbitt, T. (1995) *Latin American Society*, 2nd edn, London: Longman.
Cueva, A. (1982) *The Process of Political Domination in Ecuador*, New Brunswick, NJ: Transaction Books.
Da Matta, R. (1985) *Exploracoes: Ensaios de Sociologia Interpretativa*, Rio de Janeiro: Zahar.
—— (1990) *Carnivals, Rogues and Heroes*, Rio de Janeiro: Zahar.
Daniels, S. (1993) *Fields of Vision: Landscape Imagery and National Identity in England and the US*, Cambridge: Polity.
Davies, C. 1993 'Women and a redefinition of Argentine nationhood', *Bulletin of Latin American Research* 12 (3): 333–41.
Declaración de Quito (1990) *Encuentro Continental de Pueblos Indígenas Declaración de Quito*, Quito: (no publisher given).
Degregori, C. I., Blondet, C. and Lynch, N. (1986) *Conquistadores de un Nuevo Mundo: de Invasores a Ciudadanos en San Martin de Porras*, Lima: Instituto de Estudios Peruanos.
Deler, J. P. and Saint-Geours, Y. (comps) (1986) *Estados y Naciones en Los Andes: Hacia una Historia Comparativa*, 2 vols, Lima: Instituto de Estudios Peruanos.
Demélas, M. D. (1986) 'Una revolución conservadora de fundamento religioso: Ecuador (1809–1875)', in Deler, J. and Saint Geours, Y. (comps) *Estados y Naciones en Los Andes*, vol. 2, Lima: Instituto de Estudios Peruanos.
Demélas, M. D. and Saint-Geours, Y. (1986) 'Una revolución conservadora de fundamento religioso: el Ecuador (1809–1875)', in Deler, J. and Saint-Geours, Y. (comps) *Estados y Naciones en Los Andes: Hacia una Historia Comparativa*, Lima: Instituto de Estudios Peruanos.
Derrida, J. (1977) *Of Grammatology*, Baltimore, MD: Johns Hopkins University Press.
—— (1978) *Writing and Difference*, London: Routledge.
Desmond, J. (1993) 'Where is "the nation"? Public discourse, the body and visual display', *East–West Film Journal* 7 (2): 81–110.
Deutsch, K. (1953) *Nationalism and Social Communication: An Inquiry into the Foundations of Nationalism*, Cambridge, MA: Harvard University Press.
Díaz-Polanco, A. (1989) 'Etnias y democracia nacional en América Latina', *América Indigena* XLIX (1): 35–53.
Dickenson, J. (1992) *The Past is a Foreign Country: A Case Study of Minas Gerais, Brazil*, Liverpool papers in Human Geography new series 3, Liverpool University.
—— (1994) 'Nostalgia for a gilded past? Museums in Minas Gerais, Brazil', in Kaplan, F. (ed.) *Museums and the Making of 'Ourselves'*, London: Leicester University Press.
Diprose, R. and Ferrell, R. (eds) (1991) *Cartographies*, London: Allen & Unwin.
Dodds, K. J. (1993) 'Geography, identity and the creation of the Argentine state', *Bulletin of Latin American Research* 12 (3): 311–32.
Dofny, J. and Akiwowo, A. (eds) (1980) *National and Ethnic Movements*, Beverly Hills, CA: Sage.
Donald, J. (1988) 'How English is it? Popular literature and national literature', *New Formations* 2, repr. in Carter, E., Donald, J. and Squires, J. (eds) (1993) *Space and Place: Theories of Identity and Location*, London: Lawrence & Wishart.
Do Nascimento, A. (1989) *Brazil: Mixture or Massacre: Essays in the Genocide of Black People*, Dover, MA: Majority Press.
Dore, E. (1996) 'Property, marriage and sexuality in rural Nicaragua: across the gendered divide, 1830–1875', in Dore, E. (ed.) *Gender and Power in Latin America: Debates in Theory and Practice*, New York: Monthly Review Press.
Dore, E. (ed.) (1996) *Gender and Power in Latin America, Debates in Theory and Practice*, New York: Monthly Review Press.
Dorfman, A. and Mattelart, A (1975) *How to Read Donald Duck: Imperialist Ideology in the Disney Comic*, New York: Basic Books.
Ecuador (n.d.) *Código Penal del Ecuador*, Quito: Republic of Ecuador.
Ecuador (1984) *National Plan*, Quito: Republic of Ecuador.
Ecuador (1990) *Censos Nacionales 25 de Noviembre 1990*, Quito: Republic of Ecuador.
Ecuador (1991) *V Censo de Poblacion y IV de Vivienda 1990: Resumen Nacional*, Quito, INEC.

Ecuador (1992) *Plan y programas de estudio de tercero, cuarto, quinto y sexto años de la escuela general básico*, Quito: PROMECEB.

Ecuador (1993) *Constitución Política de la República del Ecuador Interpretaciones*, Quito: Republic of Ecuador.

Elson, D. (ed.) (1992) *Male Bias in the Development Process*, Manchester: Manchester University Press.

Enloe, C. (1989) *Bananas, Beaches and Bases: Making Feminist Sense of International Politics*, Berkeley: University of California Press.

Escobar, A. (1988) 'Power and visibility: development and the invention and management of the Third World', *Cultural Anthropology* 3(4): 428–43.

—— (1995) *Encountering Development: The Making and the Un-making of the Third World*, Boulder, CO: Westview Press.

Escobar, E. and Alvarez, S. (eds) (1992) *The Making of Social Movements in Latin America*, Boulder, CO: Westview Press.

Escolar, M., Quintero, S. and Reboratti, C. (1994) 'Geographical identity and patriotic representation in Argentina', in Hooson, D. (ed.) *Geography and National Identity*, Oxford: Blackwell.

Esteva, G. (1987) 'Regenerating people's space', *Alternatives* 12 (1): 125–52.

Estupiñán Bass, N. (1992) *Curfew*, trans. H. Richards, Washington, DC: Afro-Hispanic Institute.

Fabian, J. (1983) *Time and the Other: How Anthropology Makes its Object*, New York: Columbia University Press.

Federación Interprovincial de los Centros Shuar (1986) 'Resoluciones de la XXIII Asamblea General de la Federación', *América Indígena* XLVI dic.: 99–102.

Ferguson, J. and Gupta, A. (eds) (1992) 'Space, identity and the politics of difference', *Cultural Anthropology* 7 (1): 1–22.

Fisher, J. (1993) *Out of the Shadows: Women, Resistance and Politics in South America*, London: Latin American Bureau.

Foord, J. and Gregson, N. (1986) 'Patriarchy: towards a reconceptualisation', *Antipode* 18(2): 186–211.

Foucault, M. (1964) 'Le langage de l'éspace', *Critique* 203: 378–82.

—— (1977) *Discipline and Punish: The Birth of the Prison*, London: Allen Lane.

Fox, G. (ed.) (1990) *Nationalist Ideologies and the Production of National Cultures*, American Ethnological Society Monograph Series, no. 2, Washington, DC: American Anthropological Association.

Freston, P (1994) 'Popular protestants in Brazilian politics: a novel turn in sect–state relations', *Social Compass* 41 (4): 537–70.

Freyre, G. (1986a) *The Masters and the Slaves: A Study in the Development of Brazilian Civilisation*, trans. S. Putnam, Berkeley: University of California Press (originally published 1933).

—— (1986b) *The Mansions and the Shanties*, trans. H. Onís, Berkeley: University of California Press (originally published 1936).

—— (1986c) *Order and Progress: Brazil from Monarch to Republic*, trans. R. W. Horton, Berkeley: University of California Press (originally published 1959).

Friedman, J. (1992) 'Myth, history and political identity', *Cultural Anthropology* 7 (2): 194–210.

Fukuyama, F. (1993) *The End of History and the Last Man*, Harmondsworth: Penguin.

Fundación El Comercio (1993) *Ecuador y Perú: Futuro de Paz?*, Quito: Fundación El Comercio.

Galeano, E. (1973) *Open Veins of Latin America: Five Centuries of the Pillage of a Continent*, trans. C. Belfrage, New York: Monthly Review Press.

García Gonzalez, L. (1992) *Geografía, Historia y Cívica: Primer Curso, Ciclo Básico*, 13th edn, Quito: Editora Andina. Also volumes for *Segundo Curso, Ciclo Básico* and *Tercer Curso, Ciclo Básico*.

Garrard-Burnett, V. and Stoll, D. (eds) (1993) *Rethinking Protestantism in Latin America*, Philadelphia, PA: Temple University Press.

Geertz, C. (1973) *The Interpretation of Cultures*, New York: Basic Books.

Gellner, E. (1983) *Nations and Nationalism*, Oxford: Blackwell.

—— (1994) *Encounters with Nationalism*, Oxford: Blackwell.

Giddens, A. (1990) *The Consequences of Modernity*, Cambridge: Polity.

—— (1991) *Modernity and Self-identity*, Cambridge: Polity.

Gillis, J. (1994) 'Memory and identity: the history of a relationship', in Gillis, J. (ed.) *Commemorations: The Politics of National Identity*, Princeton, NJ: Princeton University Press.

Gilman, S. (1991) *The Jew's Body*, New York/London: Routledge.

Gilroy, P. (1993) *Small Acts: Thoughts on the Politics of Black Cultures*, London: Serpent's Tail.

Goffin, A. M. (1994) *The Rise of Protestant Evangelism in Ecuador, 1895–1990*, Florida: University Press of Florida.

Goldin, L. and Metz, B. (1991) 'An expression of cultural change: invisible converts to protestantism among highland Guatemala Mayas', *Ethnology* XXX (4): 325–38.

Gomez–Quiñones, J. (1982) 'The national question, self determination and nationalism', *Latin American Perspectives* 9 (2): 62–83.

Gott, R. (1995) 'The new reformation', *Guardian* 10 June.

Graham, R. (ed) (1990) *The Idea of Race in Latin America 1870–1940*, Austin: University of Texas Press.

Gramsci, A. (1971) *Selections from the Prison Notebooks*, trans. Q. Hoare and G. Nowell-Smith, London: Lawrence & Wishart.

Green, D. (1991) *Faces of Latin America*, London: Latin American Bureau.

Greenfield, L. (1992) *Nationalism: Five Roads to Modernity*, Cambridge, MA: Harvard University Press.

Grefa, V. (1993) 'Principales problemas de la región amazónica desde la perspectiva de la CONFENAIE', in Ruiz, L. (ed.) *Amazonía: Escenarios y Conflictos*, Quito: CEDIME.

Gregory, D. (1994) *Geographical Imaginations*, Oxford: Blackwell.

Gruffud, P. (1990) *Reach for the Sky: the Air and English Cultural Nationalism*, Nottingham: Working paper, Department of Geography, University of Nottingham.

—— (1991) 'Guest editorial – landscape, heritage and national identity', *Landscape Research* 16: 1–2.

Guerrero Arias, P. (1993) *El Saber del Mundo de los Condores: Identidad e Insurgencia de la Cultura Andina*, Quito: Abya—Yala.

Gupta, A. (1992) 'The song of the nonaligned world: transnational identities and the reinscription of space in late capitalism', *Cultural Anthropology* 7 (1): 63–79.

Gupta, A. and Ferguson, J. (1992) 'Beyond culture: space, identity and the politics of difference', *Cultural Anthropology* 7 (1): 6–23.

Gutierrez, G. (1992) 'Search for identity', *Latin American Perspectives* 19 (3): 61–6.

Gutierrez, N. (1990) 'Memoria indígena en el nacionalismo precursor de México y Perú el siglo XVIII', *Estudios Interdisciplinarios de América Latina y el Caribe*, 99–111.

Guy, D. (1991) *Sex and Danger in Buenos Aires: Prostitution, Family and Nation in Argentina*, Lincoln: University of Nebraska Press.

Hall, S. (1990) 'Cultural identity and diaspora', in Rutherford, J. (ed.) *Identity: Community, Culture, Difference*, London: Lawrence & Wishart.

—— (1991) 'The local and the global: globalization and ethnicity', in King, A. (ed.) *Globalization and the World System*, London: Macmillan.

Hanchard, M. A. (1994) *Orpheus and Power: The Movimento Negro of Rio de Janeiro and São Paulo, Brazil, 1945–1988*, Princeton, NJ: Princeton University Press.

Handler, R. (1994) 'Is "identity" a useful cross-cultural concept?', in Gillis, J. (ed.) *Commemorations: The Politics of National Identity*, Princeton, NJ: Princeton University Press.

Harley, J. B. (1988) 'Maps, knowledge and power', in D. Cosgrove and S. Daniels (eds) *The Iconography of Landscape*, Cambridge: Cambridge University Press.

Harrison, R. (1989) *Signs, Songs and Memory in the Andes*, Austin: Texas University Press.

Harvey, D. (1990) *The Condition of Post Modernity: An Enquiry into the Origins of Cultural Change*, Oxford: Basil Blackwell.

Hendrikson, C. (1991) 'Images of the Indian in Guatemala: the role of indigenous dress in Indian and Latino constructions', in Urban, G. and Scherzer J. (eds) *Nation-States and Indians in Latin America*, Austin: University of Texas Press.

Henriquez, J., Holloway, W., Urwin, C., Venn, C. and Walkerdine, V. (1984) *Changing the Subject*, London: Routledge.

Hepple, L. (1991) 'Metaphor, geopolitical discourse and the military in South America', in Duncan, J. and Barnes, T. (eds) *Writing Worlds*, London: Routledge.

Hernandez-Díaz, J. (1994) 'National identity and indigenous ethnicity in Mexico', *Canadian Review of Studies in Nationalism* XXI (1–2): 71–81.

Hidalgo, A. (1986) 'Presentación', in Vera, H. (ed.) *Mitad del Mundo Lat. 0 0'0": 250 Años del Sistema Métrico*, Quito: GAF.

History Workshop Journal (1992) Special issue on Latin America.

Hobsbawm, E. (1983) 'Introduction: Inventing traditions', in Hobsbawm, E. and Roger, T. (eds) *The Invention of Tradition*, Cambridge: Cambridge University Press.

—— (1990) *Nations and Nationalism Since 1780: Programme, Myth, Reality*, Cambridge: Cambridge University Press.

Hobsbawm, E. and Ranger, T. (eds) (1983) *The Invention of Tradition*, Cambridge: Cambridge University Press.

Hooson, D. (ed.) (1994) *Geography and National Identity*, Oxford: Blackwell.

Hudelson, J. E. (1987) *La Cultura Quichua de Transición: su Expansión y Desarrollo en el Alto Amazonas*, Quito: Abya-Yala.

Hurtado, O. (1977) *El Poder Político en el Ecuador*, Barcelona: Ariel.

IAEN (Instituto de Altos Estudios Nacionales) (1994) *Plan General de Estudios: XXI Curso Superior de Seguridad Nacional y Desarrollo*, Quito: IAEN.

Ibarra, A. (1992) *Los Indígenas y el Estado en el Ecuador: la Práctica Neo-Indigenista*, Quito: Abya-Yala.

INPC (Instituto Nacional de Patrimonio Cultural) (1989) *Legislación Nacional y Textos Internacionales Sobre la Protección del Patrimonio Cultural*, Quito: INPC.

Iturralde, D. (1981) 'Nacionalidades étnicas y políticas culturales en Ecuador', *América Indígena* XLI (3): 387–97.

Jackson, P. (1989) *Maps of Meaning*, London: Unwin Hyman.

Jacobs, J. (1995) 'Kooramindanjie: place and the postcolonial', *History Workshop Journal* 39: 165–81.

—— (1996) *Edge of Empire: Postcolonialism and the City*, London: Routledge.

Jameson, F. (1984) 'Postmodernism and the logic of late capitalism', *New Left Review* 146: 79–99.

—— (1986) 'Third World literature in the era of multinational capitalism', *Social Text* 15: 59–80.

Jaquette, J. (ed.) (1989) *The Women's Movement in Latin America: Feminism and the Transition to Democracy*, London: Unwin Hyman.

Jarosz, L. (1992) 'Constructing the dark continent: metaphor as geographic representation of Africa', *Geografiska Annaler* 74 (B): 105–15.

Jayawardena, K. (1988) *Feminism and Nationalism in the Third World*, London: Zed Press.

Jelin, E. (ed.) (1990) *Women and Social Change in Latin America*, London: Zed/UNRISD.

Jha, H. (1981) 'Stages of nationalism: some hypothetical considerations', in Dofny, J. and Akiwowo, A. (eds) *National and Ethnic Movements*, New York: Sage.

Johnson, N. (1995) 'Cast in stone: monuments, geography and nationalism', *Society and Space* 13 (1): 51–66.

Kabeer, N. (1994) *Reversed Realities: Gender Hierarchies in Development Thought*, London: Verso.

Kandiyoti, D. (1991a) 'Identity and its discontents: women and the nation', *Millenium: Journal of International Studies* 20(3): 429–43.

—— (1991b) *Women, Islam and the State*, London: Macmillan.

Kaplan, F. (ed.) (1994) *Museums and the Making of 'Ourselves': The Role of Objects in National Identity*, London: Leicester University Press.

Karakras, A. (1992) *Los Nacionalidades Indias y el Estado Ecuatoriano*, Quito: Editorial Tincui–CONAIE.

Katzenberger, E. (1995) *First World Ha Ha Ha! The Zapatista Challenge*, San Francisco, CA: City Lights Books.

Kedourie, E. (1960) *Nationalism*, Oxford: Blackwell.

Keith, M. and Pile, S. (eds) (1993) *Place and the Politics of Identity*, London: Routledge.

Kellas, J. C. (1991) *The Politics of Nationalism and Ethnicity*, London: Macmillan.

Kimberling, J. with FCUNE (1993) *Crudo Amazónico*, Quito: EDUC.

King, J. (1990) *Magical Reels: A History of Cinema in Latin America*, London: Verso.

Klump, K. (1970) 'Black traders of north highland Ecuador', in Whitten, N. and Szwed, J. (eds) *AfroAmerican Anthropology: Contemporary Perspectives*, Glencoe, IL: Free Press.

Knapp, G. (1991) *Geografía Quichua de la Sierrra del Ecuador*, 3rd edn, Quito: Abya-Yala.

Kuper, S. (1994) *Football Against the Enemy*, London: Orion.

Kuppers, G. (1994) *Compañeras: Voices from the Latin America Women's Movement*, London: Latin American Bureau.

Laclau, E. (1977) *Politics and Ideology in Marxist Theory*, London: New Left Books.

—— (1990) *New Reflections of the Revolution of Our Time*, London: Verso.

—— (ed.) (1994) *The Making of Political Identities*, London: Verso.

Laclau, E. and Mouffe, C. (1985) *Hegemony and Socialist Strategy: Towards A Radical Democratic Politics*, London: Verso.

Laclau, E. and Zac, A. (1994) 'Minding the gap: the subject of politics', in Laclau, E. (ed.) (1994) *The Making of Political Identities*, London: Verso.

Lalive d'Epinay, C. (1975) *Religion, Dynamique Social et Dépendance: Les Mouvements Protestants en Argentine et en Chili*, Paris: Mouton.

Lancaster, R. N. (1991) 'Skin colour, race and racism in Nicaragua', *Ethnology* 30 (4): 339–54.

Larrain, J. (1989) *Theories of Development*, Oxford: Blackwell.

—— (1991) *Ideology and Cultural Identity: Modernity and the Third World Presence*, Cambridge: Polity.

Lash, S. and Friedman, J. (eds) (1992) *Modernity and Identity*, Oxford: Blackwell.

Latin American Subaltern Studies Group (1993) 'Founding statement', in Beverley, J. and Oveido, J. (eds) *The Postmodernism Debate in Latin America*, special issue of *Boundary 2*, 20 (3): 110–21.

Laurie, N. (1995) 'Negotiating gender: women and emergency employment in Peru', unpublished PhD dissertation, University of London.

Lefebvre, H. (1991) *The Production of Space*, Oxford: Blackwell.

Lehmann, D. (1991) *Democracy and Development in Latin America*, Cambridge: Polity.

Lever, J. (1986) *Soccer Madness*, Chicago: University of Chicago Press.

Little, P. (1994) 'Identidades amaznicas e identidades de colonos', in Ruiz, L. (ed.) *Amazonia: Escenarios y Conflictos*, Quito: CEDIME.

Llorens, J. A. (1991) 'Andean voices on Lima Airwaves: highland migrants and radio broadcasting in Peru', *Studies in Latin American Popular Culture* 10: 177–90.

Logan, K. (1988) 'Women's political activity and empowerment in Latin America urban movements', in Gmelch, G. and Zenner, W. (eds) *Urban Life*, Prospect Heights, NJ: Waveland Press.

Lomnitz-Adler, C. (1992) *Exits from the Labyrinth: Culture and Ideology in Mexican National Space*, Berkeley: University of California Press.

Loomba, A. (1993) 'Over worlding the "Third World"', in Williams, P. and Chrisham, L. (eds) *Colonial Discourse and Post-Colonial Theory*, Brighton: Harvester.

López Vigil, J. I. (1995) *Rebel Radio: The Story of El Salvador's Radio Venceremos*, trans. M. Fried, London: Latin American Bureau.

Lowell Lewis, J. (1992) *Ring of Liberation: Deceptive Discourse in Brazilian Capoeira*, Chicago: University of Chicago Press.

Luna, L. (1993) 'Movimientos de mujeres y participación política en América Latina', Cali, Colombia: Paper prepared for Seminario Internacional Presente y Futuro de los Estudios de Género en América Latina, November.

Lutz, C. A. and Collins, J. L. (1993) *Reading National Geographic*, Chicago/London: University of Chicago Press.

Lyotard, J.-F. (1984) *The Postmodern Condition: A Report of Knowledge*, Manchester: Manchester University Press.

McCallum, C. (1994) 'A nation of whores, a government of bastards: perspectives on Brazilian nationalism from an urban heartland', unpublished manuscript.

McClintock, A. (1993a) 'Family feuds: gender, nationalism and the family', *Feminist Review* 44: 61–80.

—— (1993b) 'The angel of progress: pitfalls in the term "Post-Colonialism"', in Williams, P. and Chrisham, L. (eds) *Colonial Discourse and Post-Colonial Theory*, Brighton: Harvester.

McClure, K. (1992) 'On the subject of rights: pluralism, plurality and political identity', in Mouffe, C. (ed.) *Dimensions of Radical Democracy, Pluralism Citizenship, Community*, London: Verso.

McGee Deutsch, S. (1991) 'Gender and sociopolitical change in twentieth-century Latin America', *Hispanic American Historical Review* 71 (2): 259.

Machado Vieira, L. M. (1993) 'We learned to think politically: the influence of the Catholic Church and the feminist movement on the emergence of the health movement of the Jardim Nordeste Area in São Paulo, Brazil', in Radcliffe, S. and Westwood, S. (eds) *'Viva': Women and Popular Protest in Latin America*, London: Routledge.

MacRae, E. (1992) 'Homosexual identities in transitional Brazilian politics', in Escobar, E. and Alvarez, S. (eds) *The Making of Social Movements in Latin America*, Boulder, CO: Westview Press.

Malkki, L. (1992) 'National Geographic: the rooting of peoples and the territorialization of national identity among scholars and refugees', *Cultural Anthropology* 7(1): 24–44.

Mallon, F. (1995) *Peasant and Nation: The Making of Postcolonial Mexico and Peru*, Berkeley: University of California Press.

Marston, S. (1990) 'Who are the "people"?: gender, citizenship and the making of the American nation', *Society and Space* 8: 449–58.

Martin, B. and Mohanty, C. T. (1986) 'Feminist politics: what's home got to do with it?', in de Lauretis, T. (ed.) *Feminist Studies/Cultural Studies*, London: Macmillan.

Martin, D. (1990) *Tongues of Fire: The Explosion of Protestantism in Latin America*, Oxford: Blackwell.

Martín-Barbero, J. (1993) *Communication, Culture and Hegemony: From Media to Meditations*, London: Sage.

Martín-Barbero, J. and Muñoz, S. (1992) *Televisión Y Melodrama*, Bogotá: Tercer Mundo Editores.

Martz, J. M. (1985) 'Ecuador: the right takes command', *Current History* 84 February.

Masiello, F. (1992) *Between Civilization and Barbarism: Women, Nation and Literary Culture in Modern Argentina*, Lincoln, NB London: University of Nebraska Press.

Mason, T. (1995) *Passion of the People? Football in South America*, London: Verso.

Massey, D. (1993) 'Power geometry and a progressive sense of place', in Bird, J., Curtis, B., Putnam, T., Robertson, G. and Tickner, L. (eds) (1993) *Mapping the Futures*, London: Routledge.

—— (1994) *Space, Place and Gender*, Cambridge: Polity.

—— (1995) 'Thinking radical democracy spatially', *Environment and Planning D: Society and Space* 13: 283–8.

Mattelart, A. (1979) *Multinational Corporations and the Control of Culture: Ideological Apparatuses of Imperialism*, Hassocks: Harvester.

Maynard, K. (1993) 'Protestant theories and anthropological knowledge: convergent models in the Ecuadorean Sierra', *Cultural Anthropology* 8 (2): 246–67.

Melcon, R. and Smith, S. (eds) (1961) *The Real Madrid Book of Football*, London: World Distributors.

Melucci, A. (1989) *Nomads of the Present: Social Movements and Individual Needs in Contemporary Society*, London: Hutchinson.

—— (1992) 'Liberation or meaning? Social movements, culture and democracy', in Nederveen Pieterse, J. (ed.) *Emancipations, Modern and Postmodern*, London: Sage.

Menchú, R. (1983) *I . . . Rigoberta Menchú: An Indian Woman in Guatemala*, trans. A. Wright, London: Verso.

Menendez Carrión, A. (1988) 'Mujér y participación política en Ecuador: elementos para la configuración de una temática', Quito: FLACSO.

Metz, A. (1992) 'Leopoldo Lugones and the Jews: the contradictions of Argentine nationalism', *Ethnic and Racial Studies* 15 (1): 36–60.

Miles, A. (1994) 'Helping out at home: gender socialisation, moral development and devil stories in Cuenca, Ecuador', *Ethnos* 22 (2): 132–57.

Miles, R. (1989) *Racism*, London: Routledge.

Miller, R. (1993) *Britain and Latin America in the 19th and 20th Centuries*, London: Longman.

Minguet, C. (1973) 'El concepto de nación, pueblo, estado y patria en las generaciones de la Independencia', in Aymes, J. (ed.) *Recherches sur le Monde Hispanique au Dix-Neuvieme Siecle*, Lille: Université de Lille III.

Mohanty, C. (1991) 'Cartographies of struggle: third world women and the politics of feminism', in Mohanty, C., Russo, A. and Torres, L. (eds) *Third World Women and the Politics of Feminism*, Bloomington: Indiana University Press.

Molina Flores, A. (1994) *Las Fuerzas Armadas Ecuatorianas: Paz y Desarrollo*, Quito: Asociación Latinoamericano para los Derechos Humanos.

Moncayo, P. (1977) *Ecuador: Grietas en la Dominación*, Quito: Universidad.

Moore, H. (1994) *A Passion for Difference*, Cambridge: Polity.

Moore, S. F. (1989) 'The production of cultural pluralism as a process', *Public Culture* 1 (2): 26–48.

Morales-Moreno, L.G. (1994) 'History and patriotism in the national Museum of Mexico', in Kaplan, F. (ed.) *Museums and the Making of 'Ourselves'*, London: Leicester University Press.

Morley, D. and Robins, K. (1995) *Spaces of Identity: Global Media, Electronic Landscapes and Cultural Boundaries*, London: Routledge.

Moser, C. O. N. (1993a) 'Adjustment from below: Low-income women, time and the triple role in Guayaquil, Ecuador', in Radcliffe, S. and Westwood, S. (eds) *'Viva': Women and Popular Protest in Latin America*, London: Routledge.

—— (1993b) *Gender Planning and Development: Theory, Practice and Training*, London: Routledge.

Mosse, G. (1975) *Nationalization of the Masses*, New York: Fertig.

Mouffe, C. (1992) *Dimensions of Radical Democracy, Pluralism, Citizenship, Community*, London: Verso.

—— (1993) *The Return of the Political*, London: Verso.

—— (1995) 'Post-Marxism, democracy and identity', *Environment and Planning D: Society and Space* 13: 259–65.

Mundo Shuar (1983) *El Indígena y la Tierra Conferencia de Ginebra September 1981*, Quito: Mundo Shuar.

Naranjo, M., Pereira, J. and Whitten, N. E., (1984) *Temas Sobre la Continuidad y Adaptación Cultural Ecuatoriana*, Quito: Universidad Católica.

Nash, C. (1993) 'Remapping and renaming: new cartographies of identity, gender and landscape in Ireland', *Feminist Review* 44: 39–57.

Nederveen Pieterse, J. (ed.) (1992) *Emancipations, Modern and Postmodern*, London: Sage.

Norwood, J. and Monk, J. (eds) (1987) *The Desert is No Lady: Southwestern Landscapes in Women's Writing and Art*, New Haven, CT: Yale University Press.

Orlove, B. (1986) 'Tomar la bandera: política y trago en el sur peruano', in Briggs, L., Lianque

Chana. D. and Platt. D. (comps) *Identidades Andinas y Lógicas del Compesinado*. Lima: Mosca Azul.
—— (1993) 'Putting race in its place: order in colonial and postcolonial Peruvian geography'. *Social Research* 60 (2): 301–36.
Ortiz. A. (1987) 'Graves conflictos en la Amazonia ecuatoriana'. *América Indígena* XLVIII Dic: 153–6.
Paerregaard. K. (1994) 'Conversion, migration and social identity'. *Ethnos*. 59 (3–4): 168–86.
Palmer. S. (1993) 'Getting to know the unknown soldier: official nationalism in Costa Rica 1880–1900'. *Journal of Latin American Studies* 25 (1): 45–72.
Palomeque. S. (1990) *Cuenca en el Siglo XIX: la Articulción de una Región*. Quito: Abya-Yala.
Parker. A.. Russo. M.. Sommer. D. and Yaeger. P. (eds) (1992) *Nationalisms and Sexualities*. London: Routledge.
Parker. R. G. (1991) *Bodies. Pleasures and Passions: Sexual Culture in Contemporary Brazil*. Boston, MA: Beacon Press.
Pateman. C. (1988) *The Sexual Contract*. Cambridge: Polity.
Perelli. C. (1994) 'Memoria de sangre: fear, hope and disenchantment in Argentina'. in Boyarin. J. (ed.) *Remapping Memory: The Politics of Time–Space*. Minneapolis/ London: University of Minnesota Press.
Pietri. A.-L. (1987) 'La provincia de Loja en el conjunto nacional ecuatoriana'. in *Coloquio Estado y Región en Los Andes*. Cuzco: Centro de Estudios Rurales Andinos Bartolomé de las Casas.
Pion-Berlin. D. (1991) 'Between confrontation and accommodation: military and government policy in democratic Argentina'. *Journal of Latin American Studies* 23: 543–71.
Platt. T. (1982) *Estado Boliviano y Ayllu Andino: Tierra. y Tributo en el Norte de Potosí*. Lima: Instituto de Estudios Peruanos.
—— (1992) 'Writing. Shamanism and identity: or voices from Abya-Yala'. *History Workshop Journal* 34: 132–47.
—— (1993) 'Simón Bolívar. The Sun of Justice and the Amerindian virgin: Andean conceptions of the Patria in 19th century Potosí'. *Journal of Latin American Studies* 25 (1): 159–86.
Poole. R. (1992) 'On National Identity'. *Radical Philosophy* 62: 14–19.
Portocarrero. G. and Oliart. P. (1989) *El Perú Desde la Escuela*. Lima: Instituto de Apoyo Agrario.
Pratt. M. L. (1990) 'Woman. literature and national brotherhood'. in Bergmann. E. (ed.) *Women, Culture and Politics in Latin America*. Berkeley: University of California.
—— (1992) *Imperial Eyes: Travel Writing and Transculturation*. London: Routledge.
Probyn. E. (1991) 'Travels in the postmodern: making sense of the local'. in Nicholson. L. (ed.) *Feminism and Postmodernism*. London: Routledge.
Quintero Lopez. L. (1987) 'El estado terrateniente en el Ecuador (1809–1895)'. in Delers. J. P. and Saint-Geours. Y. (eds) *Estados y Naciones en Los Andes*. Lima: Instituto de Estudios Peruanos.
Quintero. L. and Silva. E. (1991) *Ecuador: Una Nación en Ciernes*. Quito: FLACSO.
Race and Class (1992) special issue 'The curse of Columbus' 33 (3) Jan–Mar.
Radcliffe. S. A. (1990a) 'Marking the boundaries between the community. the state and history in the Andes'. *Journal of Latin American Studies* 22 (3): 575–94.
—— (1990b) 'Ethnicity. patriarchy and incorporation into the nation: female migrants as domestic servants in Peru'. *Society and Space* 8: 379–93.
—— (1990c) 'Multiple identities and negotiation over gender: female peasant leaders in Peru'. *Bulletin of Latin American Research* 9 (2): 229–47.
—— (1993a) '"Women have to rise up – like the great women fighters": the state and peasant women in Peru'. in Radcliffe. S. and Westwood. S. (eds) *'Viva': Women and Popular Protest in Latin America*. London: Routledge.
—— (1993b) 'Women's place/el lugar de mujeres: Latin America and the politics of gender identity'. in Keith. M. and Pile. S. (eds) *Place and the Politics of Identity*. London: Routledge.
—— (1996a) 'Imaginative geographies. post-colonialism and national identities: contemporary discourses of the nation in Ecuador'. *Ecumene* 3 (1): 23–42.
—— (1996b) 'Gendered nations: nostalgia. development and territory in Ecuador'. *Gender, Place and Culture* 3 (1): 5–21.
Radcliffe. S. and Westwood. S. (eds) *Viva: Women and Popular Protest in Latin America*. London: Routledge.
Ramón V. G. (1990) *El Poder y Los Norandinos*. Quito: CAAP.
Rappaport. J. (1984) 'Las misiones protestantes y la resistencia indígena en el sur de Columbia'. *América Indigena* 44 (1): 111–27.
—— (1992) 'Fictive foundations: national romances and subaltern ethnicity in Latin America'. *History Workshop Journal* 34: 119–31.
Rattansi. A. (1992) 'Changing the subject? Racism. culture and education'. in Rattansi. A. and Donald. J. (eds) *'Race'. Culture and Difference*. Milton Keynes: Open University Press.

—— (1994) '"Western" racisms, ethnicities and identities in a "postmodern" frame', in Rattansi, A. and Westwood, S. (eds) *Racism, Modernity and Identity*, Cambridge: Polity.

Rattansi, A. and Westwood, S. (eds) (1994) *Racism, Modernity and Identity: On the Western Front*, Cambridge: Polity.

Ree, J. (1992) 'Internationality', *Radical Philosophy* 60: 3–11.

Reid Andrews, G. (1992) 'Black political protest in Sao Paulo 1888–1988', *Journal of Latin American Studies* 24 (1): 147–72.

Renan, E. (1990) 'What is a Nation?', in Bhabha, H. (ed.) *Nation and Narration*, New York: Routledge.

Restrepo, M. (1993) 'El problema de la frontera en la construcción del espacio amazónico', in Ruiz, L. (ed.) *Amazonia: Escenarios y Conflictos*, Quito: CEDIME.

Reynolds, D. (1994) 'Political geography: the power of place and spaciality of politics', *Progress in Human Geography* 18 (2): 234–47.

Rival, L. (1994) 'State schools against forest life: the impact of formal education on the Huaorani of Amazonian Ecuador', in Alsop, T. and Brock, T. (eds) *Oxford Educational Series*, Oxford: Oxford University Press.

Robinson, D. (1989) 'The language and significance of place in Latin America', in Agnew, J. and Duncan, J. (eds) *The Power of Place*, Boston, MA: Unwin Hyman.

Rose, G. (1993) *Feminism and Geography: The Limits of Geographical Knowledge*, Cambridge: Polity.

Ross, A. (ed.) (1988) *Universal Abandon? The Politics of Postmodernism*, Minneapolis: University of Minnesota Press.

Rouquié, A. (1987) *The Military and the State in Latin America*, Berkeley: University of California Press.

Rouse, R., Ferguson, J. and Gupta, A. (eds) (1992) *Culture, Power, Place: Exploration in Critical Anthropology*, Boulder, CO: Westview Press.

Rowe, W. and Schelling, V. (1991) *Memory and Modernity: Popular Culture in Latin America*, London: Verso.

Ruiz, L. (ed.) (1994) *Amazonía: Escenarios y conflictos*, Quito: CEDIME.

Rutherford, J. (ed.) (1990) *Identity: Community, Culture and Difference*, London: Lawrence & Wishart.

Said, E. (1978) *Orientalism*, London: Routledge.

—— (1994) *Culture and Imperialism*, London: Vintage.

Saint-Geours, Y. (1984) 'La sierra du Nord et du centre en Equateur. 1830–1875', *Bulletin des Institut Francais des Etudes Andines* 13 (1–2): 1–15.

—— (1989) 'La genèse de L'industrie en Equateur 1860–1914', *Bulletin de l'Institut Français des Etudes Andines* 13 (3–4): 21–8.

Salazar, E. (1981) 'The Federación Shuar and the colonization frontier', in Whitten, N. E. (ed.) *Cultural Transformations and Ethnicity in Modern Ecuador*, Urbana: University of Illinois Press.

Salazar, J. (1993) 'Black poetry of Coastal Ecuador', in Kleymeyer, C. (ed.) *Cultural Expression and Grassroots Development: Cases from Latin America and the Caribbean*, Boulder, CO: Lynne Reinner.

Sallnow, M. (1987) *Pilgrims of the Andes: Regional Cults in Cuzco*, Washington, DC: Smithsonian Institute Press.

Samuel, R. (ed.) (1989) *Patriotism: The Making and Unmaking of British National Identity*, London: Routledge.

Saporta Sternbach, N., Navarro-Aranguren, M., Chuchryk, P. and Alvarez, S. (1992) 'Feminisms in Latin America: from Bogotá to San Bernardo', in Escobar, A. and Alvarez, S. (eds) *The Making of Social Movements in Latin America*, Boulder, CO: Westview Press.

Scarpaci, J. L. and Frazier, L. J. (1993) 'State terror: ideology, protest and the gendering of landscapes', *Progress in Human Geography* 17 (1): 1–21.

Schama, S. (1995) *Landscape and Memory*, London: HarperCollins.

Schirmer, J. (1993) 'The seeking of truth and the gendering of consciousness: Comadres of El Salvador and CONAVIGUA of Guatemala', in Radcliffe, S. and Westwood, S. (eds) *'Viva': Women and Popular Protest in Latin America*, London: Routledge.

Schlesinger, P. (1987) 'On national identity: some conceptions and misconceptions criticized', *Social Science Information* 26: 219–64.

Scott, C. (1995) *Gender and Development: Rethinking Modernization and Dependency Theory*, Boulder, CO: Lynne Reinner.

Scott, J. (1986) 'Gender: a useful category of historical analysis', *American Historical Review* 91 (5): 1067–70.

Segal, D. and Handler, R. (1994) 'Introduction', in Segal, D. and Handler, R. (eds) Issue on 'Nations, colonies and metropoles', *Social Analysis* 33.

Seidman, S. (1994) *Contested Knowledge: Social Theory in the Postmodern Era*, Oxford: Blackwell.

Selverston, M. (1992) 'Politicized ethnicity and the Nation State in Ecuador', Latin American Studies Association (LASA) paper, Los Angeles: LASA.

Serrano, F. (1994) 'Las organizaciones indígenas de la Amazonía ecuatoriana', in Ruiz, L. (ed.) *Amazonía: Escenarios y conflictos*, Quito: CEDIME.

Sharp, J. (1993) 'Publishing American identity: popular geopolitics, myth and The Reader's Digest', *Political Geography* 12 (6): 491–503.

Sharp, W. F. (1976) *Slavery on the Spanish Frontier: The Colombian Choco 1680–1810*, Norman: University of Oklahoma Press.

Sherman, D. and Rogoff, I. (eds) (1994) *Museum Culture – Histories, Discourses, Spectacles*, London: Routledge.

Sibley, D. (1995) *Geographies of Exclusion: Society and Difference in the West*, London: Routledge.

Silva, E. (1991) *Los Mitos de la Ecuatorianidad Ensayo Sobre la Identidad Nacional*, Quito: Abya-Yala.

Silverblatt, I. (1988) 'Political memories and colonising symbols: Santiago and the mountain gods of colonial Peru', in Hill, J. (ed.) *Rethinking History and Myth*, Urbana: University of Illinois Press.

Simpson, A. (1993) *Xuxa: The Mega-Marketing of Gender, Race and Modernity*, Philadelphia: Temple University Press.

Skar, S. L. (1981) 'Andean women and the concept of space/time', in S. Ardener (ed.) *Women and Space*, Oxford: Berg.

Slater, D. (1985) *New Social Movements and the State in Latin America*, Amsterdam: CEDLA.

—— (1989) *Territory and State Power in Latin America: The Peruvian Case*, London: Macmillan.

Smith, A. (1986) 'Legends and landscapes', in Smith, A., *The Ethnic Origin of Nations*, Oxford: Blackwell.

—— (1991) *National Identity*, Harmondsworth: Penguin.

Smith, L. (1992) 'Indigenous land rights in Ecuador', *Race and Class* 33 (3): 102–5.

Smith, S. J. (1990) 'Social geography: patriarchy, racism and nationalism,' *Progress in Human Geography* 19: 261–71.

Snyder, L. L. (1990) *Encyclopedia of Nationalism*, London: St James Press.

Soja, E. (1989) *Postmodern Geographies*, London: Verso.

Sommer, D. (1990) 'Irresistible romance: the foundational fictions of Latin America', in Bhabha, H. (ed.) *Nation and Narration*, London/New York: Routledge.

—— (1991) *Foundational Fictions: The National Romances of Latin America*, Berkeley/Oxford: University of California Press.

Sommers, L. K. (1991) 'Latinismo and Ethnic nationalism in cultural performance', *Studies in Latin American Popular Culture* 10: 75–86.

Stark, L. R. (1981) 'Folk models of stratification and ethnicity in the highlands of northern Ecuador', in Whitten, N. (ed.) *Cultural transformations and Ethnicity in Modern Ecuador*, Urbana: University of Illinois Press.

Stasiulis, D. and Yuval-Davis N. (eds) (1995) *Unsettling Settler Societies: Articulations of Gender, Race, Ethnicity and Class*, London: Sage.

Stepan, N. L. (1991) *"The Hour of Eugenics": Race, Gender and Nation in Latin America*, Ithaca, NY/London: Cornell University Press.

Stevens, E. (1973) 'Marianismo: the other face of *machismo* in Latin America', in Pescatello, A. (ed.) *Female and Male in Latin America*, Pittsburgh, PA: University of Pittsburgh Press.

Stolcke, V. (1988) *Coffee Planters, Workers and Wives: Class Conflict and Gender Relations on São Paulo Plantations, 1850–1980*, Basingstoke: Macmillan/St Anthony's.

Stolen, K. A. (1987) *A Media Voz: Relaciones de Género en la Sierra Ecuatoriana*, Quito: CEPLAES.

Stoll, D. (1990) *Is Latin America Turning Protestant? The Politics of Evangelical Growth*, Berkeley/Los Angeles: University of California Press.

Stubbs, J. (1995) 'In search of an unchaperoned discourse on Mariana Grajales Coello: social and political motherhood of Cuba', in Shepherd, V., Burton, B. and Bailey, B. (eds) *Engendering History: Caribbean Women in Historical Perspective*, Kingston, Jamaica: Ian Randall.

Stutzman, R. (1974) 'Black Highlanders: racism and ethnic stratification in the Ecuadorean Sierra', PhD thesis, St Louis: Washington University.

Szászdi, A. (1963) 'The historiography of the Republic of Ecuador', *American Historical Review* LXVIII (887): 503–50.

Taussig, M. (1987) *Shamanism, Colonialism and the Wild Man*, Chicago: Chicago University Press.

Taylor, A.-C. (n.d.) 'Les indiens de l'amazonie et la question ethnique en Equateur', Paris: mimeo-CNRS.

Taylor, P. (1989) *Political Geography*, 2nd edn, London: Longman.

Terán, F. (1990) *Geografía del Ecuador*, 13th edn, Quito: Libresa.

Thurner, M. (1995) '"Republicanos y la comunidad de Peruanos": unimagined political communities in postcolonial Andean Peru', *Journal of Latin American Studies* 27 (2): 291–318.

Tobar Donoso, J. (1992) *El Indio en el Ecuador Independente*, Quito: Universidad Catolica del Ecuador.

Torres, C. A. (1992) *The Church, Society and Hegemony: A Critical Sociology of Religion in Latin America*, trans. R. A. Young, London: Praeger.

Tovar, H. (1986) 'Problemas de la transición del Estado colonial al Estado nacional (1810–1850)', in Deler, J. P. and Saint-Geours, Y. (eds) *Estados y Naciones en Los Andes*, Lima: Instituto de Estudios Peruanos.

Townsend, J. G. (1995) *Women's Voices from the Rainforest*, London: Routledge.

Turino, T. (1990) 'Somos el Perú: "Cumbia Andina" and the children of Andean migrants in Lima', *Studies in Latin American Popular Culture* 9: 15–37.

—— (1991) 'The state and Andean musical production in Peru', in Urban, G. and Sherzer, J. (eds) *Nation States and Indians in Latin America*, Austin: University of Texas Press.

Uquillas, J. (1984) 'Colonization and spontaneous settlement in the Ecuadorean Amazon', in Schmink, M. and Wood, C. (eds) *Frontier Expansion in Amazonia*, Miami: University of Florida Press.

—— (1993) 'Estructuración del espacio y actividad productiva indígena en la Amazonia ecuatoriana', in Ruiz, L. (ed.) *Amazonia: Escenarios y Conflictos*, Quito: CEDIME.

Urban, G. (1991) 'The semiotics of state–indian linguistic relationships: Peru, Paraguay and Brazil', in Urban, G. and Sherzer, J. (eds) *Nation States and Indians in Latin America*, Austin: University of Texas Press.

Valentine, G. (1993) '(Hetero) sexing space: lesbian perceptions and experiences of everyday spaces', *Society and Space* 11: 395–413.

Van der Veer, P. (1994) *Religious Nationalism*, Berkeley: University of California Press.

Varese, S. (1988) 'Multiethnicity and hegemonic construction: Indian plans and the future', in Guidiere, R., Pellizi, F. and Tambiah, S. J. (eds) *Ethnicities and Nations: Processes of Interethnic Relations in Latin America, South East Asia and the Pacific*, Austin: Texas University Press.

Vargas, V. (1991) 'The women's movement in Peru: streams, spaces and knots', *European Review of Latin American and Caribbean Studies* 50: 7–50.

Véliz, C. (1992) 'Centralism and nationalism in Latin America', in Wiarda, H. J. (ed.) *Politics and Social Change in Latin America: Still a Distinct Tradition?*, Boulder, CO: Westview Press.

Vera, H. (1986) *Mitad del Mundo Lat.0.0'.0": 250 Años del Sistema Metrico*, Quito: GAF.

Verdery, K. (1993) 'Whither "nation" and "nationalism"?' *Daedalus* 122(3): 37–46.

Vila, P. (1991) 'Tango to folk: hegemony, construction and popular identities in Argentina', *Studies in Latin American Popular Culture* 10: 107–40.

Villasis Terán, E. (1992) *Elogio del Ecuador*, 2nd rev. edn, Quito: Gráficas Iberia.

de Vries, L. (1988) *Política Lingüística en Ecuador, Perú y Bolivia*, Quito: Proyecto EBI/CEDIME.

Wade, P. (1986) 'Patterns of race in Colombia', *Bulletin of Latin American Research* 5 (2): 1–19.

—— (1994) *Blackness and Race Mixture: The Dynamics of Racial Identity in Colombia*, Baltimore, MD: Johns Hopkins University Press.

Walby, S. (1990) *Theorizing Patriarchy*, Oxford: Blackwell.

—— (1992) 'Women and nation', *International Journal of Comparative Sociology* 33 (1–2): 81–100.

Walter, B. (1995) 'Irishness, gender and place', *Society and Space* 13 (1): 35–50.

Walter, L. (1981) 'Otavaleño development ethnicity and national integration', *América Indígena* XLI: 319–37.

Ware, V. (1992) *Beyond the Pale: White Women, Racism and History*, London: Verso.

Warner, M. (1985) *Monuments and Maidens*, London: Verso.

Watts, M. J. (1991) 'Mapping meaning, denoting difference, imagining identity: dialectical images and post-modern geographies', *Geografiska Annaler* 73B: 7–16.

—— (1992) 'Space for everything (a commentary)', *Cultural Anthropology* 7 (1): 115–29.

—— (1993) 'Development I: power, knowledge, discursive practice', *Progress in Human Geography* 17 (2): 257–72.

Webb Vidal, A. (1992) 'The shrimp cocktail's hidden sting', *Geographical Journal* LXIV (8).

Weismantel, M. J. (1988) *Food, Gender and Poverty in the Ecuadorean Andes*, Philadelphia: University of Pennsylvania Press.

Westwood, S. (1991) 'Racism, black masculinity and the politics of space', in Hearn, J. and Morgan, D. (eds) *Men, Masculinity and Social Theory*, London: Unwin-Heinemann.

—— (1993) 'En-gendered power: racism, nationalism and the politics of identities', Cali, Colombia: Paper prepared for Seminario Internacional Presente y Futuro de los Estudios de Género en América Latina, November.

Westwood, S. and Radcliffe, S. (1993) 'Gender, racism and the politics of identities in Latin America', in Radcliffe, S. and Westwood, S. (eds) *'Viva': Women and Popular Protest in Latin America*, London: Routledge.

Whitehead, L. (1992) 'The alternatives to "liberal democracy": a Latin American perspective', in Held, D. (ed.) *Prospects for Democracy*, Cambridge: Polity.

Whitten, N. (ed.) (1981) *Cultural Transformation and Ethnicity in Modern Ecuador*, Chicago: University of Illinois Press.

—— (1985) *Sicuanga Runa*, Chicago: University of Illinois.

Wilk, R. (1995) 'The local and the global in the political economy of beauty: from Miss Belize to Miss World', *Review of International Political Economy* 2 (1): 117–34.

Willems, E. (1967) *Followers of the New Faith: Cultural Change and the Rise of Protestantism in Brazil and Chile*, Nashville, TN: Vanderbilt University Press.

Williams, C. and Smith, A. (1983) 'The national construction of social space', *Progress in Human Geography* 7: 502–18.

Williams, R. (1987) *Towards 2000*, London: Chatto & Windus.

Williamson, E. (1992) *The Penguin History of Latin America*, Harmondsworth: Penguin.

Wilson, F. (1986) 'Conflict on a Peruvian hacienda', *Bulletin of Latin American Research* 5 (1): 65–94.

—— (1993) 'Workshops as domestic domains: reflections on small-scale industry in Mexico', *World Development* 21 (1): 67–80.

Winant, H. (1992) 'Rethinking race in Brazil', *Journal of Latin American Studies* 24(1): 173–92.

—— (1994) *Racial Conditions: Politics, Theory, Comparisons*, Minneapolis: University of Minnesota Press.

Women and Geography Study Group (WGSG) (1996) *Feminist Geographies*, London: Hutchinson.

Wright, W. (1990) *Café con Leche: Race, Class, and National Image in Venezuela*, Austin: University of Austin Press.

Young, R. J. C. (1995) *Colonial Desire: Hybridity in Theory, Culture and Race*, London: Routledge.

Yúdice, G. (1992) 'Postmodernity and transnational capitalism in Latin America', in Yúdice, G., Franco, J. and Flores, J. (eds) *On Edge: The Crisis of Contemporary Latin American Culture*, Minneapolis: University of Minnesota Press.

Yúdice, G., Franco, J. and Flores, J. (eds) (1992) *On Edge: the Crisis of Contemporary Latin American Culture*, Minneapolis: University of Minnesota Press.

Yuval-Davis, N. and Anthias, F. (1989) 'Introduction', in Yuval-Davis, N. and Anthias, F. (eds) *Women-Nation-State*, London: Macmillan.

Zamosc, L. (1994) 'Agrarian protest and the Indian movement in the Ecuadorean highlands', *Latin American Research Review* 21 (3): 37–69.

Zárate, C. G. (1993) 'Cambio ambiental y apropriación del espacio en la historia de la Alta Amazonia ecuatoriana', in Ruiz, L. (ed.) *Amazonia: Escenarios y Conflictos*, Quito: CEDIME.

Zúrita, B. (1960) 'El Instituto Geográfico Militar y los organismos internacionales', in *Instituto Geográfico Militar 1928–1960 32 Años al Servicio de la Patria*, Quito: IGM.

INDEX